797,885 Books

are available to read at

www.ForgottenBooks.com

Forgotten Books' App
Available for mobile, tablet & eReader

ISBN 978-1-333-18536-7
PIBN 10543835

This book is a reproduction of an important historical work. Forgotten Books uses state-of-the-art technology to digitally reconstruct the work, preserving the original format whilst repairing imperfections present in the aged copy. In rare cases, an imperfection in the original, such as a blemish or missing page, may be replicated in our edition. We do, however, repair the vast majority of imperfections successfully; any imperfections that remain are intentionally left to preserve the state of such historical works.

Forgotten Books is a registered trademark of FB &c Ltd.
Copyright © 2015 FB &c Ltd.
FB &c Ltd, Dalton House, 60 Windsor Avenue, London, SW19 2RR.
Company number 08720141. Registered in England and Wales.

For support please visit www.forgottenbooks.com

# 1 MONTH OF
# FREE
# READING

at

**www.ForgottenBooks.com**

By purchasing this book you are eligible for one month membership to ForgottenBooks.com, giving you unlimited access to our entire collection of over 700,000 titles via our web site and mobile apps.

To claim your free month visit:

www.forgottenbooks.com/free543835

* Offer is valid for 45 days from date of purchase. Terms and conditions apply.

**English**
**Français**
**Deutsche**
**Italiano**
**Español**
**Português**

www.forgottenbooks.com

**Mythology** Photography **Fiction**
Fishing Christianity **Art** Cooking
Essays Buddhism Freemasonry
Medicine **Biology** Music **Ancient**
**Egypt** Evolution Carpentry Physics
Dance Geology **Mathematics** Fitness
Shakespeare **Folklore** Yoga Marketing
**Confidence** Immortality Biographies
Poetry **Psychology** Witchcraft
Electronics Chemistry History **Law**
Accounting **Philosophy** Anthropology
Alchemy Drama Quantum Mechanics
Atheism Sexual Health **Ancient History**
**Entrepreneurship** Languages Sport
Paleontology Needlework Islam
**Metaphysics** Investment Archaeology
Parenting Statistics Criminology
**Motivational**

Frederic C Fowler

# BRITISH CONFERVÆ;

OR

## COLORED FIGURES AND DESCRIPTIONS

OF THE

## British Plants

REFERRED BY BOTANISTS TO THE GENUS

## CONFERVA.

By LEWIS WESTON DILLWYN, F.R.S. & F.L.S.

LONDON:
*PRINTED AND SOLD BY W. PHILLIPS,*
GEORGE YARD, LOMBARD STREET.
1809.

TO

DAWSON TURNER, Esq. A.M. F.L.S.

MEMBER OF THE IMPERIAL ACAD. NAT. CURIOSORUM,

AND OF THE

GOTTENGEN PHYSICAL SOCIETY,

THIS WORK,

AS A TOKEN OF SINCERE REGARD, AND A PUBLIC ACKNOWLEDGMENT

OF THE GREAT ASSISTANCE IT HAS RECEIVED FROM HIM,

IS RESPECTFULLY INSCRIBED.

## PREFACE TO THE FIRST FASCICULUS.

THE prefent very imperfect ftate of our knowledge of Confervæ, will, I hope, be accepted as fufficient apology for not prefacing my firft Fafciculus with any general remarks on that genus. Convinced of this deficiency, I offer the prefent work as little more than a fet of drawings, whereby the fpecies of this intricate tribe may be, in fome meafure, fixed; and which may at leaft ferve as materials for the future labours of fome more able Botanifts. I fhall add to each plate the defcription of the plant it is intended to reprefent, pointing out at the fame time whatever has ftruck me as moft remarkable in its conformation or phyfiology.—The greater part of the more minute fpecies refemble each other fo much in their natural ftate, that the microfcope alone can enable us to diftinguifh them, and therefore I have given only magnified fketches; except of thofe, in which the ftructure or ramification is fufficiently fingular to point them out, at firft fight, to the naked eye.

If the botanical world fhould approve of this undertaking, my plan is to publifh a fimilar Fafciculus every four months, by which means there will be fufficient time to examine accurately the plants I introduce. It is impoffible yet to offer a conjecture to what extent the work will reach; at present I have only examined, and that imperfectly, the environs of London, Yarmouth, and Dover, but from what I have seen in these places, I am convinced that the Confervæ are a far more numerous tribe than is in general imagined. I solicit the assistance of other Botanists, and shall receive with thanks any remarks tending to elucidate either the species previously described, or those which still remain to be introduced. I only beg leave to say, that these plants must not be judged of from dried specimens; for when the granules in their interior substance once collapse, no subsequent immersion will restore them to their former appearance, or bring back the elasticity they possessed when recent. To give as accurate an idea as possible of the relative size of each plant, it appeared best to state, after every figure, with what power the drawing was made; the numbers, therefore, denote the several magnifying powers of a common compound microscope.

Higham Lodge, Walthamstow,
June 1st. 1802.

# INTRODUCTION.

## SECT. I.

### GENERAL REMARKS.

THE Confervæ, whether confidered with regard to their external appearance, their internal ftructure, or the extraordinary manner in which the propagation of many fpecies is effected, may undoubtedly be reckoned among the moft beautiful and curious of the order of vegetables to which they belong. It was my original intention to have given in this work a magnified drawing of each Britifh fpecies, but the number of thofe already difcovered is fo great, and it is fo impoffible to obtain fpecimens of all fufficiently recent for the purpofe, that I find it a tafk almost endlefs, and above my ability to complete. I have therefore been obliged to content myfelf with giving a brief account, by way of fynopfis, of nearly all thofe fpecies which have fallen under my obfervation*, and a drawing, accompanied with a more full defcription, of moft of thofe which I have met with recent, and which have not been figured

* Since this was written, I have been induced fo far to deviate from what I had here propofed, as to give a flight fketch (generally from dried fpecimens) of all the fpecies not figured by other authors, excepting C. fanguinea, which would not revive fufficiently in water to enable me to trace its ftructure.

7387

elsewhere. In this state I offer the result of my labors to the Botanic world, in hopes that its numerous defects will be excused; when it is considered that the Confervæ were very lately involved in such obscurity as to have been publicly termed 'the opprobrium of Botany.' *

If we look back to what had previously been done in this department of science, we shall find that Linnæus was too busily engaged in the immense field he had entered on, to spare the time necessary for an investigation of the submersed Algæ, as appears both from his writings and Herbarium, in which latter scarcely any specimens of Confervæ are preserved. In the *species Plantarum*, and also in the works of most other authors, the subject is treated so slightly, that many different plants may not only be often referred to the same description, but were actually designed by the writers to be included under it; and even these short descriptions are chiefly borrowed from Dillenius, who remained almost the only original author on the *Confervæ*, till Dr. Roth published the first Fasciculus of his *Catalecta Botanica*, in 1797. Even of Hudson's descriptions in the *Flora Anglica*, many are entirely borrowed from the *Historia Muscorum*, and those which he has taken from his own observations are too short to be of much service. Lightfoot, indeed, when he relied upon himself alone, is perhaps more than any other author exempt from such a charge, and the only thing to be lamented in this excellent Botanist is, that he allowed himself so often to transcribe the works of others, who were far inferior to himself in the art either of observing or of recording their observations. Had Dillenius accustomed himself to the use of a microscope, there is little that might not have been expected from his accurate pencil; but, for the want of this assistance, he has frequently confounded several species together, which agree only in external habit, and has even described some as jointless in which dissepiments are readily observable with a common glass. The only magnified drawings of Confervæ, to which reference with any tolerable precision could be made prior to the close of the last century, were those of Mr. Ellis, in the 56th volume of

* Dr. Smith's Introductory Discourse, *Lin. Trans.* I. p. 34.

[library stamp]

the Royal Society's Tranſactions; and thoſe of Muller, in the *Flora Danica* and *Nova Acta Petropolitana;* in which works theſe diſtinguiſhed naturaliſts have diſplayed their accuſtomed accuracy and talent for minute investigation. Such being the caſe, I truſt that this work, by elucidating the ſynonymy of theſe, as well as of the more modern authors, and by the variety of new matter it contains, will be found ſo far to clear the way as to induce others, with more leiſure and ability than myſelf, to purſue the ſtudy, and perfect our knowledge of a tribe than which none will be found more intereſting. The purſuit, though not otherwiſe of high importance, tends, as Dr. Smith obſerves, 'to enliven the ſcenes of rural retirement, to relieve the mind amid the buſy purſuits of active life,' and carries with it its own reward in the conſtant ſource of amuſement which it preſents to the ſtudent wherever he goes, and in the complacency which an inveſtigation of the works of nature never fail to excite in the mind, beſides the higher object of teaching man to admire and adore his Maker in the works of his hand.

M. Girod Chantrons, in his *Recherches ſur les Conferves*, has, both by chemical analyſis, and by obſervations on their ſtructure, endeavoured to prove that the *Conferva* are either real animals or of animal origin; and that many of them are actual Polypi, others the habitations of theſe animals, and others again, aggregations of Polypi, ſo attached together as to form a tube. It appears to me, ſo far as I am able to judge from the drawings and deſcriptions, that this work is too inaccurate to merit much attention. Dr. Treviranus, in his *Biologie**, has gone ſtill further, and propoſes to unite, not only the Confervæ, but the whole claſs Cryptogamia with the Zoophytes, and thus form a fourth kingdom, intermediate between the animal and vegetable. I cannot help ſuſpecting that theſe authors have given too much ſcope to their imagination, and the more ſo, as a ſimilar analyſis by M. Vauquelin, has been attended with ſuch different reſults, as to confirm him in the oppoſite and generally

* This work, which I have not myſelf ſeen, is wholly quoted on the authority of Sprengel's *Introduction to Botany*.

received opinion. He found that the ſmall quantity of ammonia contained in *Confervæ* is combined with pyromucous acid, which is the caſe in many vegetables: that they do not give out muriate of ſoda, as Meſſrs. Chantrons and Lacroix have affirmed, but muriate and carbonate of potaſh, and if they had contained ſoda, this is only what occurs in ſeveral other plants. He conſiders the quantity of aſhes they afford as a ſtill further proof, and upon the whole entertains no doubt that their ſubſtance is truly vegetable.* M. Decandolle has alſo, in my opinion, ſucceſsfully controvered M. Chantrons' theory; and I therefore need only add, that I have never diſcovered an appearance in any of the Algæ which occaſioned the leaſt ſuſpicion in my mind that they are not true and perfect vegetables.

With regard to the preſent arrangement of the *ſubmerſed Algæ*, I have little more to add than that nothing can more fully evince our ignorance reſpecting them, or ſhew how imperfectly they have been hitherto ſtudied, than the circumſtance of ſo many diſcordant ſpecies being placed together, as thoſe of which the preſent genera are compoſed. It may probably be expected, that in a work of this kind I ſhould attempt ſome better arrangement, but, though ſatisfied of the neceſſity of ſuch a taſk, I can only lament my inadequacy to the execution of it: the time is not yet arrived; ſufficient materials are not yet collected; and it ſhould be deferred not only till the *Confervæ*, but alſo till the *Fuci*, *Ulvæ*, and *Tremellæ* are better known, as well with regard to their fructification as to the number of their ſpecies; for many are ſtill frequently diſcovered, differing eſſentially in their modes of propagation from thoſe before known. Crude and undigeſted attempts at reformation ſerve in Botany, as in other matters, to perplex rather than to enlighten, and I will therefore add nothing further on this ſubject, than that I fully agree with my friend Mr. Turner, that, previouſly to any permanent ſyſtem being eſtabliſhed, it will be neceſſary to reduce the preſent genera into one maſs, and proceed in nearly the ſame manner as if nothing had been done before.

* Journal de Phyſique, liv. p. 427.

The *Conferva* have hitherto been confidered as principally diftinguifhed from other *Algæ* by the jointed ftructure of the filaments; but this circumftance is not of itfelf fufficient to feparate them from many of the *Fuci*, nor even perhaps from fome of the *Lichens* and *Fungi*.* There are alfo feveral plants which have by the general confent of Botanifts been always called *Confervæ*, in which no joints are obfervable, fo that, if, in addition to what is here obferved, be added the many remarks upon the fame fubject that occur in the prefent work, and in Mr. Turner's admirable *Hiftory of the Fuci*, (particularly in his defcription of *F. dafyphyllus*) it appears fufficiently proved that the jointed ftructure can no longer be ufed as a diftinctive generic mark. Indeed the Confervæ muft be regarded rather as a natural family, comprehending many genera of plants than as a fingle genus, and I have therefore felt it would be abfurd, as well as unneceffary, to attempt fuch a generic character as would comprife the whole, becaufe, according to the rules of botanical philofophy, this fhould be formed from the fructification, and the fructification of the *Conferva* differs fo infinitely in different fpecies, that it would be impoffible to include them all under any fuch defcription. I have, however, for the prefent, retained *Conferva* as a general name for all thofe plants which have been, or which if known, would have been fo called by preceding authors, in the fame manner as the term Lichen was applied by Dr. Acharius, in the Prodromus of his *Lichenographia Suecica*. To thefe I have alfo added the *Byffi filamentofæ*, as they differ in no refpect from the *Confervæ* in ftructure; and fince the publication of my defcription of *C. aurea*, the propriety of this union has been eftablifhed by a difcovery of its capfules, which refemble thofe of Dr. Roth's *Ceramia*.

Drs. Ingenhouz and Girtanner, from the general prevalence of *Confervæ* in almoft all waters and moift places, have been led to fuppofe that they are generated fpontaneoufly from the decompofition of water by the folar rays; but

* Since I publifhed the defcription of *C. atro-virens*, Mr. Hooker has afcertained that it is *Cornicularia pubescens* of Acharius, but the capfules which I difcovered in July, 1806, near Beddgellart, prove that it belongs to the *Confervæ*. *Fibrillaria ramosissima* of Sowerby's Englifh Fungi, as well as fome other fpecies of the fame genus, and of the *Auriculariæ*, are links which connect the latter tribe with the Confervæ.

" omnia ex ovo" is now ſo univerſally received as an axiom, that few naturaliſts will be likely to accede to their opinion. In ſome *Confervæ*, indeed, no mode of propagation has hitherto been diſcovered, except by an elongation or expanſion, and viviparous diviſion of the filament; but analogy induces me to ſuſpect that even theſe are alſo propagated by ſeeds, as has been aſcertained to be the caſe in moſt of the other ſpecies.

Of the extent of this tribe I feel myſelf unable to offer a conjecture: more than two hundred different ſpecies have been already aſcertained in the few parts of Europe in which the *Confervæ* have been at all examined, and I have no doubt but that even our own Iſlands will be found to produce a ſtill larger number. In my *Synopſis* I have been obliged to omit ſeveral ſpecies, with ſpecimens of which I have been favored by my friends, becauſe the latter are ſo imperfect, or have ſuffered ſo much change from drying, that it is impoſſible to obtain their diſtinguiſhing characters. In addition to the ſpecies which have fallen under my own obſervation, or of which ſuch drawings or deſcriptions have been publiſhed as to leave no doubt of their identity, I have admitted only thoſe of which I poſſeſs either ſketches or ſpecimens, ſufficiently perfect to afford a tolerably correct idea of the recent plant. The accuracy, however, of all deſcriptions of *Confervæ*, which are taken from dried ſpecimens, for reaſons aſſigned in the preface to my firſt Faſciculus, may be doubted, and I therefore, whenever this has been the caſe, have prefixed an aſteriſk to the name of the ſpecies, in order that a proper allowance may be made.

The *Confervæ*, by the large quantity of oxygen that they give out, have been thought to render the air about ſtagnant waters more wholeſome; but of their uſe and economy no more is yet known than of their number. Many ſpecies remain of whoſe whole phyſiology we are intirely ignorant, and perhaps no other tribe can be found which ſtill offers ſo wide a field for diſcovery.

Of thoſe who have attempted a diviſion of the Confervæ into Genera, Dr. Roth and M. Vaucher are the authors who deſerve particular attention, and I ſhall now proceed to give a ſketch of their different arrangements.

## SECTION II.

### SYSTEM OF ROTH.

Dr. Roth has divided the ſubmerſed Algæ into the following Genera: *Fucus*, *Ceramium*, *Batrachoſpermum*, *Conferva*, *Mertenſia*, *Hydrodictyon*, *Ulva*, *Rivularia*, *Linkia*, and *Tremella*. I ſhall give the outline of each of theſe, and offer a few remarks on thoſe that contain any of the plants uſually denominated Confervæ.

Fucus.—*Veſiculæ aggregatæ, ſubſtantiæ frondis immerſæ, poris mucifluis præditæ.*

This genus, of which the definition is as vague and unmeaning as the ſame number of words can well be, is intended to comprehend a part only of the plants uſually called by the name of *Fuci*, the remainder having been referred to the following:

Ceramium.—*Filo membranaceo-cartilaginea, capſulis granuliferis ipſis adnatis.*

In this genus are made two diviſions; the firſt, '*filis conformibus*,' contains ſome of the more ſlender Fuci, and of the unjointed capſuliferous Confervæ: the other, '*filis ſpurie geniculatis*,' comprehends the jointed Fuci, and the remainder of the capſuliferous Confervæ. There is undoubtedly a great ſimilarity in the fructification of the capſuliferous ſpecies, and yet ſeveral natural tribes, if not really diſtinct families, may be perceived among them, although, as has been already obſerved, the imperfect ſtate of our knowledge, would render it imprudent to attempt at preſent to define their reſpective limits. I ſhall however enumerate thoſe which appear moſt ſtriking.

The unjointed ſpecies probably all belong to the genus *Vaucheria*, as will be hereafter mentioned.

*C. elongata*, as is obſerved in the deſcription of that ſpecies, has two kinds of capſules, ſimilar to thoſe of *Fucus ſubfuſcus* and *F. pinaſtroides*, from which it cannot be ſeparated without violence; and to theſe may probably be united the black marine, and thoſe other ſpecies in which the filament is an aggregation of ſeveral ſmaller tubes; in theſe the capſules are ovate, reticulated, and ſeſſile.

In C. *ciliata, diaphana, rubra*, and ſome others obviouſly ſimilar in character, the capſules are ovate, ſolitary, and ſubtended by two or more calyciform proceſſes.

In C. *plumula, roſea, Turneri*, and their congeners with pinnated filaments, the capſules are globoſe, numerous, and neither reticulated nor ſubtended by the abovementioned proceſſes. The capſules of C. *littoralis, pennata, ſcoparia* and *tomentoſa*, are nearly ſimilar to the foregoing, but, inſtead of being placed on the ramuli only, at the end of almoſt every joint, they are ſcattered without order on the filaments.

The capſules of C. *ſpongioſa*, and alſo moſt probably of C. *verticillata*, are oblong, petiolated, and unuſually ſmall for the ſize of the plant.

C. *ſetacea, barbata*, and thoſe which have their ſeeds imbedded in mucus, and guarded by an involucrum inſtead of a capſule, form a very diſtinct and beautiful family, which cannot be arranged with propriety in any of Dr. Roth's genera. From the deſcription in Mr. Turner's work, it appears that *F. plumoſus*, as well as ſome other Fuci, may probably be found to belong to the ſame genus, which I had hoped to have ſeen at ſome future time, when the ſubmerſed algæ ſhall be remodelled, diſtinguiſhed with the name of Mr. Borrer, to whoſe unwearied application we are indebted for our knowledge of many of its ſpecies, but I have juſt found that Dr. Acharius has, in his new *Lichenographia Univerſalis*, ſo called one of the new genera of his favorite family.

BATRACHOSPERMUM.—*Baccae polyſpermæ, coloratæ, filamenta geniculata, cartilagineo-membranacea.*

In this genus two ſpecies only are enumerated, with one of which, *B. dichotomum*, I am entirely unacquainted. The other is C. *gelatinoſa*, of which a deſcription may be found in the body of this work.

CONFERVA.—*Tubuli vel filamenta herbacea internis parietibus fructificationum granulis adspersa.*

This genus, ftill more than Ceramium, comprifes plants belonging to feveral natural families, perfectly diftinct from each other, and the learned Doctor feems to have ufed it as a receptacle for all thofe fpecies of which the fructification is unknown, or which he could not otherwife difpofe of. They are arranged under two feparate heads, "*Tubulofæ*," and "*Filamentofæ*." Of thefe, the former is compofed of the tubular *Ulvæ*, the union of which with the *Confervæ* appears to me to be by no means warranted by what is at prefent known on the fubject. The latter has three main divifions, depending on what are here termed *genicula*, but which are in the courfe of this work (perhaps improperly) named *Diffepiments*. I have ufed this word to exprefs every fort of divifion between the *vefícles* or *articuli* of thefe plants. It would certainly be defirable to diftinguifh, by different names, the different natures of thefe divifions, but they are often fo ambiguous, and in plants fo minute it is neceffarily fo difficult to examine them accurately, that I have not ventured to undertake the talk.

The only Britifh fpecies arranged in the firft divifion, "*Conformes feu continuæ*," are C. *feneftralis*, *Ulva plumofa*, and C. *dilatata*, refpecting which I muft be allowed to remark, that this arrangement of C. *feneftralis* is erroneous, as diffepiments may be obferved in the filaments when examined with the higher powers of a microfcope, while *U. plumofa* is at leaft a plant of doubtful place in the fyftem, and C. *dilatata* is the fame with my C. *vefícata*, and fhould therefore in this arrangement have been placed with *C. amphibia*, among the *Ceramia*.

The fecond divifion is entitled "*Articulatæ, geniculis fpuriis*." Dr. Roth calls thofe diffepiments *fpurious*, which have their origin in the internal ftructure, and not in the fibres which conftitute the filament. This fection is itfelf thrice fubdivided.

The firft fubdivifion is termed "*Sporangiorum annulis*." In thefe plants, Dr.

Roth ſuppoſes that the joints are in fact a feries of annular ſeed-caſes, not attached to, but diſpoſed within the filaments, at regular ſhort intervals from each other; and that theſe intervals conſtitute the ſuppoſed diſſepiments. This ſubdiviſion comprifes Vaucher's natural genus *Oſcillatoria*, and is the ſame with the ſection B. a. of my ſynopſis, but I have not uſed the word *ſporangium*, becauſe it cannot be properly applied to theſe joints, as will be hereafter ſhewn.

Of the ſecond ſubdiviſion, which is entitled, "*Utriculis matricalibus*," Dr. Roth ſays, that this ſpecies of ſpurious partition differs from thoſe formed by the *annular ſeed-caſes* above deſcribed, in this particular, that they are not viſible in the earlieſt, but only at ſome advanced ſtage of their growth, or in conſequence of ſome violent concuſſion, and that the joints can never vary from the poſition allotted to them. Whilſt the plants are young, or till their organization has been diſturbed, the internal veſicles are contiguous to each other at their extremities, and the filaments then appear in every reſpect equal and continuous; but, when at length theſe veſicles become contracted, an empty pellucid ſpace is left at each extremity, without any appearance of a true diſſepiment in the middle. The kind of joint here deſcribed is found, according to Dr. Roth, in many of the *Ceramia*, as well as in *C. ægagrophila*, *ericetorum*, and other Confervæ without any natural affinity; and it appears to me evident that the term *utriculus matricalis* cannot with propriety be uſed to define a ſpecies of joint which occurs ſo frequently in capſuliferous ſpecies.

The plants of the third ſection, "*Stricturis*," are deſtitute of real joints, but divided by annular ſtrictures at uncertain diſtances from each other. C. *toruloſa* is the only Britiſh ſpecies here arranged, unleſs my ſuſpicion ſhould prove well founded, that Dr. Roth's C. *reptans* is *Fucus opuntia*.

We next come to the third main diviſion of the filamentous Confervæ, "*Articulatæ geniculis veris*." Such alone are admitted by Dr. Roth to be true diſſepiments as actually interſect the interior of the tube, being formed by the branching of the parallel fibres, of which, together with a cellular membrane, the filament itſelf is compoſed. This ſection has four ſubdiviſions, of which the firſt is the

"*Fasciatæ.*" In these Dr. Roth is of opinion that the dissepiment does not extend wholly across the tube, but leaves it pervious for its whole length. This subdivision comprises the genus *Conjugata* of Vaucher, with *Conferva equisetifolia*, *crispata*, *ebenea*, *vivipara*, *fucicola*, and many others, and it is therefore obviously far from natural.

Of the "*Torulosæ*," which form the second subdivision, the dissepiments rise above the surface of the tube in the form of annular excrescences. Dr. Roth here supposes the dissepiments to be interwoven with a large portion of the cellular membrane, which makes them less able to resist the elasticity of the enclosed air, and they thereby become distended. C. *fluviatilis* is the only British species that occurs in this subdivision, but my observations have tended to confirm the opinion of the late lamented Dr. Mohr, that these protuberances are of a different nature, and ought not to be regarded as dissepiments.

C. *atra* is arranged by itself, and forms a third subdivision with the name of "*Institiæ*," the meaning affixed by the Doctor to which term is, that the longitudinal fibres of the filaments on attaining to the length prescribed for each joint, suddenly unite in a single point, and are bent inwards towards the cavity of the tube, thus forming an appearance similar to that of the *tortulosæ*, though in reality of different structure; and hence each joint is narrow at its origin, and gradually incrassated upwards.

The fourth and last subdivision is composed of the "*Verticillatæ*," distinguished chiefly by their verticillated, or rather imbricated, ramuli. In this C. *vertillata*, and *spongiosa* are arranged, together with some foreign species probably of the same family, and with C. *villosa*, a plant widely different both in its nature and structure.

Having taken this cursory view of the genus *Conferva*, as established by Dr. Roth, it remains only to add, that enough has already been discovered of the fructification of many of the species classed under it, to shew the necessity of their removal to other genera, and it seems to me that those only should be retained which are propagated by seeds formed within the joints, without the aſſiſtance of any external proceſs whatſoever, thus excluding even the *Conjugatæ*, which it was his intention to admit.

MERTENSIA.—*Tubuli ſub coriacei, intus articulati, ſporulæ in tunica, papillas veſicales clavatas faſciculatas efficiente, ſparſæ.*

This genus has but one ſpecies, the *Ulva lumbricalis* of Linnæus, a native of the Cape of Good Hope, with which I am totally unacquainted. According to Dr. Roth's account, the ſtructure of this plant is extremely curious. The filaments are lined on the inſide with a fine cellular membrane, and at ſhort, but equal and regular, diſtances a circle of ſpine-like proceſſes iſſue internally, over which the cellular membrane is ſpread, ſo as to cloſe up the tube tranſverſely at every diſſepiment.

HYDRODICTYON.—*Fila ſub membranacea, tubuloſa, ad angulos varios in utriculum retiformem finibus ſuis combinata, demum utriculum matri ſimilem invaginatum producentia.*

C. *reticulata* differs ſo entirely from every other known Confervæ, that Dr. Roth has very properly formed it into a ſeparate genus, with the preſent expreſſive name. The eſſential character in the ſecond Faſciculus of his Catalecta is taken wholly from the ſingular contexture of its filaments, but this he has been enabled to amend in the third, by M. Vaucher's important diſcovery of its ſtill more ſingular propagation.

ULVA.—*Membrana expanſa, diaphana, fructificationum granulis praprimis circa marginem innatis.*

The genus, as here conſtituted, is intended to comprehend only thoſe ſpecies which are compoſed of a ſingle leaf-like membrane, the tubular ones having, as before mentioned, been removed to the *Confervæ*, and *U. incraſſata*, *U. rubens*, and their affinities, to the following family.

RIVULARIA.—*Subſtantia gelatinoſo-cartilaginea, hyalina, integumento membranaceo deſtituta; fructificationes in filis geniculatis intra ſubſtantiam nidulantibus.*

The plants which Dr. Roth has referred to this genus, are ſo cloſely allied with the *Batrachosperma*, and the latter approach some other families of the *Conferva* by so many points, as to render the attempt to separate them extremely difficult. The gelatinous nature and appearance of the filaments is not sufficient, nor are the fine transparent processes into which their ramuli are drawn, for these may be also observed in *C. protensa*, *vivipara*, and some other *Conferva*. Although this affinity is so strong, yet as none of the *Rivularia* have ever been published under the latter name, I have not thought it necessary to notice them in my general synopsis.* Dr. Smith, milled by some apparent resemblance in their structure, has published some of the *Fuci*, and even *Tremella*, under this name, but I trust it will not be found necessary to retain them, or the *Rivularia* can no longer be regarded as a natural family.

LINCKIA.—*Substantia gelatinosa, hyalina, integumento membranaceo hyalino induta, farcta fructificationum granulis in lineas curvatas moniliformes ordinatis.*

Micheli used *Linckia* as a generic name for those *Tremella* with which he was acquainted, and in these he observed that the granules were arranged in regular lines. Dr. Roth, however, carrying his researches farther, discovered that in some only of the *Tremella* the seeds are thus arranged, but that in others they are scattered throughout the internal mucus without apparent order. He has therefore separated the former from *Tremella*, and with them constituted the present genus, retaining Micheli's original name. Five species have been ascertained, consisting of *Tremella nostoc* and *utriculata*; two recently discovered species, and *Ulva pruniformis*, which had been before removed to the *Tremella* by Mr. Woodward. The plant which Dr. Roth supposes to be *U. pruniformis* of Linnæus, is however essentially different from that figured in *English Botany* (t. 968,) with the same name, and which Mr. Hooker informs me is *Rivularia angulosa* of the *Catalecta Botanica*. *Linckia pruniformis* is not known to be a British species.

* I have never seen a recent specimen, but I presume from the description, that *C. echinulata* of Eng. Bot. t. 1378, belongs to this family.

Tremella.—*Subſtantia uniformis gelatinoſa, hyalina, integumento membranaceo induta, fructificationum granulis in membranæ contextu fibroſo abſque ordine ſparſis.*

Such of the plants as have been uſually called *Tremellæ*, and have the ſeeds ſcattered without order throughout the internal mucus, conſtitute, as is above obſerved, this genus.

Byssus.—In the third volume of the *Flora Germanica*, this genus is retained with the following definition, "*Filamenta vel fibrae tenuia, membranacea, lanuginoſa, extus fructificatione granulis adſperſa,*" and compriſes all the ſpecies of Hudſon's ſection, "*filamentoſæ,*" but in the ſecond volume of the *Catalecta Botanica*, ſome of them had been referred to the *Fungi*, and the "*Pulverulentæ*" had been ſeveral years before removed to the *Lichens* in the firſt volume of the *Flora*. In the third Faſciculus of the *Catalecta*, the genus *Byſſus* is not noticed, but three of the filamentous ſpecies are incorporated with the *Conferva*. I have examined *B. phoſphorea, æruginosa, velutina, purpurea, nigra, aurea*, and *ſulphurea*.* and cannot find that they poſſeſs any character to diſtinguiſh them from the *Conferva*, but *B. ſeptica* ſeems to be of a different nature, and to belong to the Fungi, with which it has been arranged by M. Perſoon.

It ſeemed neceſſary to ſtate my reaſons for not having followed the arrangement of ſo eminent a Botaniſt as Dr. Roth, and I have conſequently been obliged to point out what have appeared to me to be its leading imperfections. Every word written by ſuch an obſerver muſt, however, be of value, and although his arrangement has, in my opinion, been premature, I am convinced that a better will never be effected without a liberal uſe of his numerous obſervations.

* *Byssus sulphurea, Lichenis facie, tenuissima ac densissimè, filtrum & pannum laneum texturo similans.* Micheli, p. 211 a 17. Dill. *Hist. Musc.* p. 7. t. 1. f. 13. This ſingular ſpecies has not been diſcovered in Britain, and I am indebted to Mr. Dryander for a ſpecimen from the Bankſian Herbarium.

## SECTION III.

### SYSTEM OF VAUCHER.

M. VAUCHER has divided the Confervæ into the fix following genera. *Ectofperma*, *Conjugata*, *Hydrodictyum*, *Polyfpermum*, *Batrachofpermum*, and *Prolifera*. In purfuance of my plan, I fhall now give an outline of his ideas of the general fructification of each of thefe, with which I fhall incorporate the few obfervations I have myfelf made on the fruit of their refpective fpecies.

The ECTOSPERMES are thus defined, "*Les organs fécondans font exterieurs, et les grains font portées fur des pedoncles qui partent d'une tube ramifié.*"

The generic name *Ectofperma*, has been changed by M. Decandolle to *Vaucheria*, and the genus is with this alteration adopted in *Englifh Botany* with the following effential character, "*Antheræ, awl-fhaped, incurved. Capfules adjoining to the Antheræ, ovate, fingle feeded, in pairs or folitary.*"

M. Vaucher has traced the growth of thefe plants, through all their ftages, and fatisfactorily proved what the obfervations of Mr. Borrer have fince fully confirmed, that they are propagated by the germination of their granules. Although, in his fpecific defcriptions, M. Vaucher has called thefe granules naked feeds, yet his remarks, added to my own, induce me rather to believe that the *grains* of all the *Vaucheriæ* are monofpermous capfules, as Dr. Smith has defcribed them. I wifh, however, rather to fubmit this as a matter for future obfervation, than to exprefs a decided opinion upon the fubject, though having circulated a theory founded on a contrary opinion among my Botanical friends, I think it neceffary to fhew how I was led into this error. The fpecies figured at T. 74, under Muller's name of C. *veficata*, is *Vaucheria feffilis*, but I could not then difcover any antheræ, and as Vaucher's *grains* are reprefented to be naked feeds, and as he had not mentioned the capfules or bladder-like veficles which abounded in my fpecimens, I concluded that it muft be a different fpecies.

I afterwards discovered the grains and antheræ of *V. geminata* on some filaments, precisely resembling those of *Conferva vesicata*, but the grains were so much smaller, that, relying on M. Vaucher's description, I concluded they were naked seeds. I therefore imagined that in the former specimen the male and female organs were concealed within the capsule, and in the specimens which I afterwards gathered, that the capsules had fallen off or died away, and thus left the seeds sitting on their receptacle with the anthera exposed to view.

M. Vaucher has not been able to prove the nature of what he has called *anthera* with equal satisfaction to himself; "Cependant je ne suis pas aussi certain des fonctions auquelles est appellée la corne qui les accompagne; elle est à la verité constamment placée dans les voisinage des grains; on la voit bien repandre sa poussiere dans *l'Ectosperme ovoide* en particulier, cela est incontestible. Cependant j'ai toujours désiré quelque experience directe, qui me put convaincre de l'usage de cette corne." It appears from this quotation that M. Vaucher has been rather too hasty in his application of the term *anthera* in the specific descriptions, and that he has fallen into the common error of supposing that the analogy between phænogamous and aquatic cryptogamous plants must be perfect, without making a proper allowance for the difference that must necessarily exist in the latter from the difference of their situation. If his conjecture should be confirmed by future observation, I am of opinion that the awl-shaped processes subtended by the capsules of several *Ceramia*, and the tribe intended to have been called *Borreria*, will also prove to be male organs, and effect the fecundation of the seed in the same manner.

Of this genus, M. Vaucher has enumerated eleven species, few of which can be at all distinguished from each other except by the fructification, and this varies so much with respect to the size, number, and disposition of the capsules in almost every different mass, and even in the same specimen, that it can hardly be considered a sufficient indication of specific difference. My friend, Mr. Hooker, says he has seen petioles bearing two and some three capsules, and other capsules single and sessile on the same plant. In my description of C.

*veſicata*, (*V. ſeſſilis*) I have remarked its cloſe affinity with C. *amphibia*, and in the third Faſciculus of the *Catalecta Botanica*, Dr. Roth has arranged all the *Vaucheriæ* as mere varieties of this ſpecies. My obſervations have made me incline to this opinion with reſpect to a majority of the ſpecies, and I much doubt whether they may not be all referred to either C. *amphibia*, *dichotoma*, or *Dillwynii*. As C. *myochrous* and C. *comoides*, in their ſtructure, approach theſe ſpecies, it is poſſible that their fructification, when diſcovered, will prove ſimilar. Should this conjecture be well founded, the unjointed ſpecies form a family, ſufficiently diſtinct from the other Confervæ; and whenever the algæ are new modelled, will, I truſt, be continued with the generic name by which M. Decandolle has ſo properly diſtinguiſhed the *Ectoſpermes*.

M. Vaucher has made but few references to the works of preceding authors, and, to prevent confuſion, it muſt be remarked that theſe few are extremely inaccurate: thus all the *Ectoſpermes* are ſaid to have been compriſed by Linnæus under the name of C. *fontinalis*, with which plant none of them have the leaſt affinity. Muller's C. *veſicata* is referred to *Prolifera veſicata*, which is a widely different ſpecies, and the reference to C. *velutina* with the ſynonyma of Micheli and Dillenius is equally erroneous.

CONJUGATA.—*Les grains ſont interieurs et renfermées une à une dans des tubes cloiſonnées et toujours ſimples.*

This natural and wonderful family is better characterized by the name, than by this generic deſcription, under which many other plants might be arranged whoſe filaments have never been obſerved to conjugate.

Muller, although he publiſhed excellent drawings both of C. *nitida* and *jugalis*, entertained no idea that the difference between them merely ariſes from the fructification. Meſſrs. Charles and Romain Coquebert, who alſo diſcovered *nitida* in its conjugated ſtate, were equally ignorant of this circumſtance, though they advanced one ſtep further, and aſcertained that the globules formed by the union are true ſeeds which reproduce the ſpecies. M. Vaucher's intereſting memoir, publiſhed by the Philomatic Society of Paris, although full of important difcoveries on other Confervæ, merely confirms the foregoing obfervations, and contains but little new on the *Conjugatæ*.

In the fpring of 1802 I difcovered, that at a certain period of their growth, fmall tubes are protruded from the fimple filaments of *C. nitida*; that thefe unite with the fimilar tubes of other contiguous filaments; that the grains of the one being emptied into, coalefce with the grains of the other filament, and thus conftitute the *C. jugalis* of the *Flora Danica*.* I alfo found that this ftrange property is not confined to this fingle fpecies, but that the *C. genuflexa* of Roth is formed in like manner by an union among the fimple filaments of Muller's *C. ferpentina*, and I traced thofe of *C. fpiralis* from a fimple to a conjugated ftate. In the fummer of the fame year, Mr. Woods found *C. bipunctata* with the filaments conjugated, and the fuppofed originality of thefe difcoveries afforded me great pleafure, being then quite ignorant of M. Vaucher's continned application to this tribe. At length the appearance of his *Hiftoire des Conferves d'eau douce*, at Geneva, in 1803, fhewed that we had arrived at a knowledge of the *Conjugatæ*, and formed nearly the fame conclufions refpecting them, almoft at the fame time, and quite independently of each other.

I have fince difcovered the feeds of *C. genuflexa*; they are large and globular, and not formed within either filament, as in *C. jugalis*, but in the connecting tube, which thereby becomes greatly diftended, as is reprefented in my fupplementary plate. M. Vaucher could not difcover the feeds of this fpecies, and of the nature of his obfervations recorded in the following paffage, I cannot form any conjecture. "Depuis le moment ou j'écrivais cette défcription, j'ai vu germer cette conjugée; elle nait d'une manière fort differente de toutes les autres: la matière ne paffe pas d'un tube a un tube voifin, mais chaque loge fournit elle même une jeune plante; le tube exterieur qui fe trouve renfermè,

* When I firft made the drawings and defcriptions of *C. nitida* and *jugalis*, I had not the leaft idea that they belonged to the fame fpecies, and it was unfortunately not till juft after my firft Fafciculus had been given to the printer, that I was fully fatisfied on this fubject. My defcription of *C. fpiralis*, and the drawing B which was afterwards added, will, however, fufficiently prove that I had even then arrived at a knowledge of this curious property.

devient une jeune Conjuguée, qui étoit toute entière contenue dans le vieux tube, comme elle même contient les plantes qui doivent se dévellopper ensuite: elle en sort par l'extremité lorsqu'elle occupe la dernière loge, ou par les cotés lorsqu'elle se trouve dans une des loges du milieu." I have sometimes seen the simple filaments of C. *genuflexa* rolled round in a serpentine form, as Muller has represented, and these have been erroneously referred by M. Vaucher to a separate species.

Each joint of the *Conjugatæ* puts forth only one connecting tube, which is sometimes on one side of the filament and sometimes on the other; so that each filament is often connected with two others. A short time after the union has been effected, the granules from one joint, gradually pass into that with which it is joined, till the former at length becomes empty and colorless. The granules of both then coalesce in the other joint, or in the connecting tube, and form a globular or oblong mass, which M. Vaucher has proved to be the true seed, and has seen it germinate and reproduce the species. No seeds appear to be ever formed but by this union of the joints of two different filaments with each other, and of these united joints only one ever produces a seed. It is therefore natural to conclude, although the contents of the joints by their appearance cannot in the least be distinguished from each other, and although in things so minute and obscure it is necessary to speak with the utmost diffidence, that one contains male and the other female powers, and that their union is essential to the propagation of the *Conjugatæ*. It might indeed reasonably have been supposed that of two conjugated filaments, the whole of the one is male and the other female, but in opposition to this it generally happens that a part of the joints give out, and a part receive granules in the same filament. I have seen three filaments connected together, and the connecting tubes of the middle one have sometimes been thrown out by the joints on one side, and sometimes on the other, and the seeds have been formed in either filament, without apparent order. Each filament must therefore be considered hermaphrodite, possessing in its different joints both male and female powers, which, as in the ſnail, can only be rendered productive by contact with the oppoſite powers of other filaments. Thus this apparently inſignificant tribe affords an unique and wonderful analogy between the reproduction of the animal and vegetable kingdoms, and is a ſtriking evidence that " the power of God is over all his works, and is ſeen to the aſtoniſhment of man in the variety of his wonders."

HYDRODICTYUM. — *Chaque articulation devient elle même une nouvelle plante qui s'étend comme un réseau.*

C. *reticulàta* was firſt ſeparated from the *Conferva* under the generic name of *Hydrodictyon* by Dr. Roth, as is already mentioned, and no other plant has been ſince diſcovered with which it can be aſſociated. Its ſurpriſing mode of propagation is mentioned in my deſcription of this ſpecies, and I ſhall therefore only repeat that we are wholly indebted to the ſcrutinizing talents of M. Vaucher for this important diſcovery.

POLYSPERMA.—*Les grains ſont répandues en très grand nombre dans l'intérieur d'un tube renflé, non tranſparent et ramifié.*

C. *fluviatilis* and *glomerata* are the only ſpecies which M. Vaucher has been able to refer to the preſent genus, to which however he ſuſpects that ſeveral others alſo belong. He obſerved that the filaments of *C. fluviatilis* are lined with minute beaded threads, which at length divide, and each bead then becomes a ſeparate granule. He thinks it probable that a part of theſe grains, although they cannot be diſtinguiſhed from the others, are male organs which die away as ſoon as they have performed their office; a conjecture that ſeems rather ingenious than probable. He however aſcertained, by a courſe of well directed experiments, that at leaſt a part of theſe granules are true ſeeds, and traced their growth from the germination till they reſembled the parent plant in all reſpects. Theſe globules, both in their connected and detached ſtate, may be readily obſerved by cutting and preſſing the filaments, and, though I have failed in my endeavours to witneſs their germination, I cannot in the leaſt doubt the accuracy of M. Vaucher's obſervations, or ſuppoſe that this ſpecies is not propagated as he deſcribes. Tufts of young ſeedlings may be alſo frequently obſerved, as he deſcribes them, iſſuing from the older filaments: theſe he attributes to the germination of ſeeds which have inſinuated themſelves from the interior to the ſubſtance of the frond, and thus grow paraſitically on their parent. M. Vaucher does not ſeem to have noticed the minute hair-like proceſſes that iſſue externally from the protuberances, between which and the beaded threads on the inſide, I have not been able to diſcover any connection. Upon the ſubject of theſe diſcoveries as to the ſtructure and fructification of the *Polyſpermæ*, though I have here quoted M. Vaucher alone, having myſelf had an opportunity of conſulting no work but his, yet I feel it incumbent upon me to ſay, that the concurring teſtimony of German Botaniſts attributes the original detection of them to the late Dr. Mohr, who appears from what is ſaid by Dr. Roth, to have given an ample account of them in a number of Schrader's Journal for 1801, of which I am not aware that there is a copy at preſent in England.

*C. fluviatilis* differs widely in habit and appearance from other Britiſh Confervæ, agreeing in its real character probably with none but *C. toruloſa*, unleſs indeed the conjecture of my friend, Mr. Turner, be well founded, that *Fucus pedunculatus*, *F. aculeatus*, C. *verrucoſa**, and C. *villoſa* may belong to the ſame tribe. With reſpect to C. *glomerata*, which has not the leaſt affinity to any of theſe, M. Vaucher ſays little more than that he found its joints contain numerous minute granules, and thence concluded they were ſeeds. Of this, although he continued his obſervations with unremitted aſſiduity for two years, he could, however, obtain no further proof, than that the ſtones in a river were covered with ſomewhat ſimilar granules, which germinated and produced this ſpecies. He therefore determined on the arrangement of C. *glomerata* in this genus, but candidly allows, 'C'eſt bien plus l'analogie et le raiſonnement que les obſervations directes, qui nous ont conduit dans les conjectures que nous avons haſardées ſur ſon Hiſtoire.'

* I have omitted this ſpecies in my ſynopſis, becauſe having carefully examined its internal ſtructure, I am decidedly of opinion that it has no claim whatever to a place among the Confervæ. It will, I hope, appear in Mr. Turner's Hiſtoria Fucorum.

BATRACHOSPERMUM.—" *Chaque Anneau, après s'être ſeparé de l'ancienne Conferve, pouſſe de toutes parts des nouvelles ramifications.*"

In this genus, *Conferva gelatinoſa* and *mutabilis*, together with the *Conferva chara*, and *Rivularia elegans* of Roth, and *Ulva incraſſata* of Hudſon, are arranged by M. Vaucher, who is of opinion, " Que cette famille fort différente des autres ſe multiplie par ſes anneaux. Lorſqu'elle a acquis à peu près tout ſon accroiſſement, les anneaux dont elle eſt composée, ſe rompent et ſe ſeparent. Le plus grand nombre d'entr'eux, ſur tout lorſque toutes les parties de la Conferve ſe detruiſent en même temp, s'éloignent de manière qu'il n'eſt plus poſſible de les ſuivre. Les autres reſtent attachés aux filets à cauſe de leur viſcoſité ; peu à peu ils croiſſent et s'étendent. La forme qu'ils ont alors n'est pas regulière, mais elle eſt aſſez ſemblable dans tous les grains. Inſenſiblement ils groſſiſſent ; en même temps ils acquirent aſſez de tranſparence pour qu'on puiſſe voir dans leur interieur la Batrachoſperme à laquelle ils doivent donner naiſſauce ; enfin l'envellope, qui les contenuit, ne pouvant plus ſe prêter à leur accroiſſement il en ſort de toutes parts un grand nombre de petites plantes, qui s'étendent en rayonnant autour d'un même point, et chaque filet eſt un tronc principal de la Conferve que ſe dévelope. Cet état de demi dévelloppement eſt celui des grains noirs que l'on aperçoit ſur la Batrachoſperme à collier (*Conferva gelatinoſa.*) Ils y ſont retenus, comme je l'ai dit, par des filets de la plante ; et ſi on les examine au microſcope, on trouve à leur centre l'anneau dont il eſt ici queſtion, qui pouſſe de toutes parts des filets rayonnans et déjà articulés."

M. Vancher, as well as Dr. Roth, has conjectured, that the delicate capillary threads which are ſeen iſſuing from the extremities of the ramuli of theſe plants, and conſtitute one of its moſt obvious characters, may be ſpermatick veſſels, but the only circumſtance which materially favors this idea is, that they fall off when the plant has attained to a certain age.

I have not been able to diſcover the fructification of any of the ſpecies arranged in this genus, except *C. gelatinoſa*. Of this I have given a drawing, and it appears to conſiſt of an aggregation of ſeeds, reſembling a compound berry, which I have ſeen germinate, both whilſt attached to and when ſeparated from the parent plant. I am ſorry to differ ſo materially from M. Vaucher on this ſubject, and I apprehend it would not have been the caſe if he had uſed a higher magnifying power.

Whether M. Vaucher has done right in uniting the *Rivulariæ* with *Batrachoſpermum*, further obſervations are in my opinion ſtill wanting to decide. I confeſs myſelf inclined to believe that the fructification of theſe genera will be found to be different, but my friends, Mr. Turner and Mr. Hooker, whoſe united opinions muſt have far greater weight, agree with M. Vaucher that they ſhould be joined.

PROLIFERA.—*Il ſort des parties renflées ou des Bourrelets du vieux tube, des filets cylindriques qui s'étendent en tout ſens, et qui après avoir pris un aſſez grand accroiſſement, ſe ſéparent enſuite de leur Mère, pour devenir eux mêmes une Conferve parfaite*

The following extract will ſerve more fully to ſhew M. Vaucher's idea of the manner in which theſe Confervæ are propagated. "Lorſque les prolifères ſont prêtes à ſe reproduire, on voit naître, le long des tubes des renflemens cylindriques, que l'on prendroit pour des nœuds, ſi la plante n'étoit pas d'ailleurs cloiſonnée. Ces Bourrelets d'abord peu ſenſibles, groſſiſſent bientôt, enſuite ils ſe couvrent d'une matière pulverulente qui eſt formée ou des débris qui flottaient dans le liquide, et qui ont été retenus par le Bourrelet; ou d'une matière qui s'eſt ſécrété de la Conferve. Lorſque cette pouſſière a ſéjourné quelque temps ſur le Bourrelet, on voit ſortir ſes nombreux filets qui forment d'abord de petites têtes arrondies. Malheureuſement cette pouſſière en même temps qu'elle ſemble favoriſer l'accroiſſement, gêne beaucoup l'obſervateur. On ne peut guères voir le premier dévelloppement de la jeune plante, et juger par exemple, ſi elle ſort de la ſurface du Bourrelet ou du centre. Quoi qu'il en ſoit, les jeunes filets s'étendent rapidement ſur toute la circonférence du Bourrelet où ils forment comme une houppe de poils. Peu à peu leur cloiſons commencent à ſe marquer, bientôt leurs tubes reſſemblent en petit à celui de la grande prolifère; enfin ils fe féparent pour allér former ailleurs un nouvel individu femblable à celui fur lequel ils ont pris naiffance; *mais j'avoue que je n'aie pas vu de féparation*, quoique je n'aie aucun lieu de douter qu'elle ne s'opère."

I have now before me the variety mentioned in my defcription of C. *rivularis**, and more fully defcribed in my fynopfis, with fhort fpine-like proceffes, refembling both in fize and fhape thofe figured on Vaucher's† *P. crifpa*. If, as M. Vaucher imagines, thefe are a proliferous progeny, it muft be fuppofed that they would refemble the parent, not only in their joints, but alfo in the fhape of the filaments; the latter are, however, cylindrical throughout, and the former, at leaft in fome fpecies, are reprefented as remarkably acuminated; and of the proceffes of C. *rivularis* at this time under my obfervation, many, not the tenth of an inch in length, are as pointed as poffible, although the diameter at their bafe equals that of the main filament. I examined this variety during a fortnight, but could not obferve that the ramuli at all encreafed in length, or fuffered any change, till at the end of that period, the whole died away, and difappeared together. M. Vaucher has not noticed how far their length bears any regular proportion to that of the main filament, and he admits, contrary to his generic definition, that he has never feen them feparate from it. In the defcription of *P. floccofa*, which is probably the fame with my C. *punctalis*, he fays, "Elle fe multiplie avec une telle rapidité qu'elle couvre au bout de quelques jours des places confidérables dans lefquelles on ne l'avolt pas d'abord aperçue," and yet he could never difcover any proliferous tendency, or any other means whatever by which this increafe was effected. I therefore wonder at this fpecies having been arranged with the *Proliferæ*, but M. Vaucher poffeffes the rare merit of never concealing or diftorting truth to ferve a favorite theory, and expreffes himfelf throughout fo doubtfully of the ufe and nature of the branch-like proceffes, that it is rather furprifing he fhould have founded the genus with fuch an uncertain character. It is neverthelefs probable, from M.

* *Prolifera rivularis* of Vaucher differs from this fpecies in its much longer joints.

† See the drawing of this fpecies in my fupplementary plate B.

Vaucher's defcription, that the proceffes which he obferved on fome fpecies are of a different nature from thofe refembling thorns above mentioned. It feems to me that the fructification of the *Proliferæ* confifts in their internal granules, and, equally with thofe of *C. fluviatilis*, it is reafonable to fuppofe, that thefe feeds may in fome inftances become lodged, and germinate in the fubftance of the filament, which germination would neceffarily occafion the frond in that part to fwell, and thus produce the *Bourrelets*, which M. Vaucher defcribes. It does not, however, appear that the filaments thus generated ever arrive at maturity, and I am decidedly of opinion that this is not the mode defigned by nature for the propagation of thefe Confervæ.

OSCILLATORIA.—This is the name given in M. Vaucher's Hiftory of the *Tremellæ*, to the Confervæ of Dr. Roth's fection '*Sporangiorum annulis*,' which are here arranged fo as to conftitute a feparate genus.

M. Vaucher obferved that *C. fontinalis*, and thofe of its congeners which float on the furface of water, are generally attended by "une efpece de feutre" "douce et onctueufe au toucher," which he fuppofes is of the fame nature as the internal mucus of the *Tremellæ*, and he compares the filaments themfelves to the beaded granules of the *Linckiæ*. I have never feen this felt-like fubftance except with C. *fontinalis*, and have always confidered it as decayed vegetable or other extraneous matter, in which the plant likes to grow, nor can I find that it bears the leaft refemblance to the internal mucus of a *Tremella*. At all events it is not fufficiently general to warrant the removal of the genus; for M. Vaucher admits that it is not met with in any of the fpecies which grow on ftones, or on other fubftances, and thefe, I believe, conftitute a majority of the genus. The fuppofed fpontaneous motion of the filaments firft noticed by M. Adanfon, and mentioned in my defcription of *C. limofa*, however, feems principally to have induced him to remove the *Ofcillatoriæ* from the *Confervæ* to the animal kingdom, for to this he fuppofes that the *Tremellæ* belong. During the laft eight years I have frequently examined feveral fpecies in hopes of difcovering this mark of animality, but muft confefs I could never obferve any motion that might not be attributed to their wonderfully rapid growth, which muſt occaſion ſuch thickly entangled filaments to preſs againſt each other; to the water in which they are examined, the ſlighteſt motion of which is ſufficient to agitate them, or to the numerous animalculæ with which they are conſtantly infeſted.

It would exceed the limits of my preſent undertaking to give a detail of M. Vaucher's numerous conjectures, and curious remarks, and I ſhall therefore now confine myſelf to the relation of what I have myſelf obſerved concerning this family.

The *Oſcillatoriæ* conſtitute a natural genus, and are diſtinguiſhable at firſt ſight by their numerous filaments ſo thickly matted together as to form a jelly-like maſs. The filaments, when examined with the higher powers of a microſcope, appear to me equally obtuſe at both ends, and are regularly divided by remarkably delicate diſſepiments into extremely ſhort joints. Some of the diſſepiments may be obſerved of a darker color and thicker ſubſtance than the others, and at theſe I believe the filaments divide into ſeparate fragments, each of which, as M. Adanſon firſt obſerved, " Devient abſolument ſemblable a celui dont il s'étoit ſéparée, et capable d'en produire a ſon tour de nouveau." In *C. vaginata*, however, the filaments are multiplied by a longitudinal inſtead of a tranſverſe diviſion, as appears in my deſcription of that ſpecies. The diameter of the filaments of this family, never varies according to their age, as in other Confervæ, but is conſtantly the ſame in every ſpecies, and hence M. Vaucher has been led to ſuppoſe that they are always propagated by the viviparous diviſion only, and never by ſeed. This opinion I was for ſome time inclined to adopt, till it was ſhaken by an appearance of capſules on ſome ſpecimens of *C. decorticans*, which is repreſented in my ſupplementary plate. They are ſo unuſually large in proportion to the thickneſs of the filament, that at firſt ſight I thought they were of the ſame nature with galls, or thoſe excreſcences that are ſo frequently inhabited by the *Cyclops* on the *Vauchériæ*, but, when I applied the higheſt powers of my microſcope, I found their ſhape too regular and well defined, and themfelves furrounded by a pellucid limbus fo entirely refembling that of many of the *Ceramia*, as to give them every appearance of true capfules. I kept the fpecimens feveral days, but could not obferve any feeds efcape from them, nor have I fince been able to difcover any thing at all fimilar in either this or any other fpecies of *Ofcillatoria*, and fubfequent difcoveries have encreafed my fufpicions that they were not capfules, and have induced me to believe that C. *decorticans*, as well as the other fpecies of this family, are propagated by feed in a different manner.

In examining fome fpecimens of *C. diftorta*, I obferved a number of detached globules of the fame color, and of about equal diameter with that of the filaments, and I alfo obferved that in fome filaments which were partly empty, the remaining joints had affumed a fimilar globular fhape. Some of the detached globules had become of an oblong form, and a diffepiment was then obfervable in the middle, while others were more elongated with four joints, and others were ftill longer, fo as to form a regular feries, beginning with the globule, and ending in a perfect filament. I have, therefore, no doubt, fo far as it is poffible to ftate any opinion on objects fo minute and obfcure without any doubt, that each joint at length becomes a feed, which efcapes at the apex of the filament, and that by its evolution the fpecies is propagated. I have obferved a precifely fimilar appearance in C. *mirabilis*, and have alfo feen detached granules, apparently of the fame nature among the filaments of *C. fontinalis* and *C. muralis*, and in both thefe fpecies, thofe filaments which are partly empty have their remaining joints of a more globular form, than they are in thofe which continue perfect.

I have afcertained that the filaments of C. *diftorta* conjugate in a fingular manner, (which, together with the fructification, is reprefented in my fupplementary plate A) and that the fuppofed ramifications of this fpecies are thus conftituted. C. *diftorta* is therefore moft clofely allied with *C. mirabilis* and *C. majufcula*, and I incline to the opinion that here, as well as in the *Conjugatæ*, an union of their filaments is in fome manner effential to their fructification.

The *Oſcillatoriæ*, beſides their general accordance in ſtructure, are connected in different points by *C. diſſiliens* and *confervicola* with the other *Confervæ*, nor can I, with all due deference to the opinion of M. Vaucher, allow myſelf to doubt that the propriety of retaining them among the ſubmerſed Algæ in the vegetable kingdom will be admitted by almoſt every Naturaliſt, and eſpecially by thoſe who make this department of Botany their ſtudy.

To conclude, although I cannot give implicit aſſent to all M. Vaucher's obſervations and deductions, yet the greater part of his phyſiological diſcoveries are ſo well eſtabliſhed and ſo important, that they form a memorable epoch in the hiſtory of the ſubmerſed Algæ. He has the credit of having firſt raiſed a Conferva from ſeed, and of having traced it through the different ſtages of its growth, and, to mention one only of his many diſcoveries, that of the wonderful propagation of *C. reticulata*, is in itſelf ſufficient to render his name reſpectable, as long as ſcientific merit continues to be held in eſteem.

---

## SECTION IV.

### SYSTEM OF DECANDOLLE.

I NOW turn to Decandolle, whoſe arrangement of the ſubmerſed Algæ in the *Flore Francoiſe*, and *Flora Gallica*, is however hardly worth notice. It is principally formed by incorporating the two foregoing ſyſtems, with much alteration but little or no improvement in their genera, as will be ſufficiently ſhewn by the following outline of his arrangement.

> NOSTOC.—*Integumentum vireſcens membranaceum intus farctum gelatina mucoſa filamentis moniliformibus intertexta.*

In M. Vaucher's ſyſtem, *Tremella* conſtitutes an order which is referred to the animal kingdom, and compriſes the two genera of *Oſcillatoria* and *Noſtoc*.

The latter contains the plants ufually called *Tremellæ*, which M. Decandolle, under Vaucher's generic name, has here reftored to their place in the vegetable kingdom.

RIVULARIA.—*Membranæ fubcartilagineæ, lobatæ aut ramofæ, muco gelatinofo obtectæ.*

This however is not Dr. Roth's genus, but rather the fame with *Ulva* of Vaucher, and contains, 1ft. his *U. gelatinofa*, under the name of *R. tubulofa*. 2d. *Ulva lubrica* of Roth. 3d. *R. fœtida*, which is probably my *C. fœtida*, and 4th, a new fpecies, with the name of *R. Halleri*.

ULVA.—*Frondes membranaceæ. Semina fub epidermide latitantia, fæpius aggregatæ, frondis ipfius deftructione exeuntia.*

This genus in addition to moft of the *Ulvæ*, comprifes all the *Ulva-like Fuci*, with *F. digitatus*, *F. bulbofus*, *F. tomentofus*, &c.

FUCUS.—*Algæ marinæ membranaceæ aut filamentofæ. Capfulæ aut femina aggregata in tuberculos nunc laterales, nunc terminales, apice poro dehifcentes.*

The genus as here conftituted comprifes a part only of the plants ufually called by the name of *Fucus*, fome having, as is above mentioned, been removed to the *Ulvæ*, and *F. pinaftroides* and *F. filum* to the following family. *Ulva plumofa*, which Dr. Roth has carried to his genus *Conferva*, is here, and with equal impropriety, defcribed under the name of *Fucus arbufcula*.

CERAMIUM.—*Stirpes filamentofæ marinæ fimplices aut ramofæ, diffepimentis tranfverfalibus nodofo-articulatæ; tubercula polyfperma fub globofa lateralia aut terminalia.*

The fpecies of Dr. Roth's fecond divifion of *Ceramium*, '*filis fpurie geniculatis*' conftitute this genus, and thofe of the firft divifion are fent back to the Fuci.

DIATOMA.—*Plantæ pfeudo-parafiticæ oculo nudo vix confpicuæ, filamentis fimplicibus articulatis, articulis in adulta planta tranfverfim fectis.*

The fpecies arranged in the fection "*articulis folutis*" of my fynopfis, conftrtute a natural family, and may be all referred to this genus, in which however only Roth's C. *mucor* and C. *flocculofa* are here enumerated.

CHANTRANSIA.—*Filamenta ramofa, diffepimentis inftructa; femina minutiffima, intra filamenta recondita, in quoque articulo numerofa.*

This genus is named by M. Decandolle in honor of M. Girod Chantrons. With one of the eight fpecies here enumerated I am entirely unacquainted, and the remaining feven may be referred to at leaft three different natural families. 1ft. C. *torulofa* and C. *fluviatilis* belong to Vaucher's *Polyfperma*, with which C. *glomerata* alfo here retained certainly poffeffes no affinity. 2d. C. *atra*, which in my opinion is undoubtedly a *Batrachofpermum*. And 3d, C. *rivularis*, C. *crifpa*, and C. *veficata* of Vaucher, which belong to his family of *Proliferæ*.

CONFERVA.—*Filamenta fimplicia, diffepimentis inftructa, interdum ope tubuli inter fe conjuncta, materia viridi, nunc fpiraliter, nunc bifttellatim, nunc fparfim difpofita intra loculos farcta. Semina in quoque loculo folitaria.*

The genus as here conftituted is the fame with *Conjugata* of Vaucher, and contains all the fpecies arranged as fuch in the *Hiftorie des Conferves d'eau douce*. *Prolifera parafitica* and *P. floccofa* of the fame author, are alfo added under a feparate fection, entitled, "*Haud planè notæ.*"

BATRACHOSPERMUM.—*Filamenta muco gelatinofo obtecta, ramofa, ramis filo hyalino plus minufae elongato terminatis; annulis ovatis, folidis, ad extremum progrediendo decrefcentibus. Corpufculis hirtis (plantularum rudimentis) inter ramos fparfis.*

This genus is the fame with Vaucher's *Batrachofpermum*, and is intended to comprehend the whole of the *Rivulariæ*, as well as the *Batrachofperma* of Roth.

HYDRODYCTION.—*Habitus faccatus, fere claufus, retiformis, interftitiis, feu areolis polygonis.*

C. *reticulata*, as in the fyftems of Roth and Vaucher, is here placed by itfelf.

VAUCHERIA.—*Filamenta herbacea diffepimentis planè deftituta. Semina externa, primo tubo adfixa, tandem caduca.*

This, as I have before remarked, is the fame with Vaucher's genus *Ectofperma*.

## SECTION V.

### CONFERVÆ OF HUDSON.

Having, through the kindnefs of Dr. Williams, had repeated accefs to the Dillenian Herbarium, and received fome valuable information refpecting the fynonymy of the Flora Anglica from Sir Thomas Frankland and the Rev. Hugh Davies, who were both intimately acquainted with its diftinguifhed author, I feel happy in being able to remove the uncertainty that has hitherto attended the elucidation of many of Hudfon's Confervæ. I fhall therefore offer a few obfervations on each of the doubtful fpecies, and refer the remainder to the corresponding figures of the prefent work.

1. C. *rivularis*. T. 39. Var. β. is *C. nitida* T. 4.

2. C. *fontinalis*. T. 64.

3. *C. violacea*. A plant which exactly agrees with Dillenius's and Lightfoot's defcriptions, and alfo with fome of the fpecimens in the Dillenian Herbarium, grows abundantly on the ftones in fome rapid rivulets in the neighbourhood of Swanfea, and feems to be only a flight variety of C. *decorticans*. Mr. Dickfon gave me a fpecimen of C. *diftorta*, gathered in the Highlands, under the name of C. *violacea*; but, although the former, efpecially when dried, is of a ftriking violet color, yet it differs entirely from the latter in its mode of growth, as defcribed by Dillenius.

4. *C. furcata*. The late Mr. Pitchford gave me an authentic fpecimen, marked by Hudfon " C. *furcata*," which is nothing but a narrow variety of C. *dichotoma* I have little doubt that Dillenius's No. 10, which Hudfon calls *furcata* β, is a variety of *C. amphibia*:—*C. amphibia* and C. *dichotoma* are, however, very clofely allied.

5. *C. dichotoma*. T. 15.

6. C. *bullofa*. I think there can be no doubt that many of the fpecies whofe

filaments grow fufficiently entangled to retain air bubbles, and are thereby floated on the furface of the water, were confounded together, and conftitute the prefent fuppofed fpecies.

7. *C. canalicularis* feems to me certainly nothing more than one of the numerous varieties of C. *amphibia*, which grows about mills and other falls of water, exactly as Dillenius has defcribed it. This opinion is confirmed by Mr. Turner's *Obfervations on the Dillenian Herbarium*, publifhed in the *Tranfactions of the Linnæan Society*.

8. *C. amphibia.* T. 41.

9. C. *rigida.* My own obfervations at Oxford confirm Mr. Turner's opinion, that this is nothing but C. *glomerata* encrufted with fome extraneous matter.

10. C. *fæniculacea.* This is a Fucus, as appears both by the Dillenian Herbarium, and by a fpecimen which Mr. Davies received from Hudfon. By calling it a Fucus I do not mean to exprefs any opinion upon its fructification, which is at prefent unknown, but merely to fay that it is quite deftitute of joints.

11. *C. littoralis.* T. 31.

12. C. *tomentofa.* T. 32.

13. C. *albida.* This plant, which has long been wholly unknown to Botanifts, appears from a very careful examination of the Dillenian fpecimen, by Mr. Hooker, Mr. Turner, and myfelf, under the microfcope, to be really a diftinct fpecies, and is fo defcribed in my fynopfis, and figured in the fupplementary plate E. Mr. H. Davies has obligingly favored me with a plant which had been fo named by Hudfon, and which is the C. *Hookeri* of this work.

14. C. *æruginofa* is defcribed in my fynopfis from the Dillenian fpecimen, of which a fragment is alfo reprefented in the fupplementary plate E.

15. *C. nigra.* Authentic fpecimens in the Herbaria of Sir Thomas Frankland, and the Rev. H. Davies, prove that my *C. atro-rubefcens* is this fpecies.

16. *C. fcoparia.* T. 52.

17. C. *cancellata* is *Sertularia fpinofa.*

18. *C. multifida.* This, as well as C. *imbricata*, on the authority of an authentic ſpecimen ſent by the Biſhop of Carliſle to Mr. Turner, appears clearly to belong to the C. *equiſetifolia* of this work, T. 54; I have, however, retained the name of C. *multifida* to the plant ſo called in *Engliſh Botany*, as the name, though then erroneouſly applied, is really applicable to the plant, and, not being attached to any other, may fairly be left to it.

19. C. *ſpongioſa.* T. 42.

20. C. *reticulata.* T. 97.

21. *C. fluviatilis.* T. 29.

22. C. *atra.* T. 11.

23. C. *gelatinoſa.* T. 32.

24. C. *capillaris.* T. 9.

25. C. *corallina.* T. 98.

26. *C. ſetacea.* T. 82.

27. C. *elongata.* T. 33.

28. C. *ciliata.* T. 53.

29. C. *polymorpha.* T. 44.

30. C. *tubuloſa.* The ſpecimen according with Hudſon's reference, in the Dillenian Herbarium, ſeems, as Mr. Turner remarked, to be only an unuſually thick variety of C. *rubra*, and I have myſelf gathered nearly ſimilar appearances of this ever varying ſpecies.

31. *C. rubra.* T. 34.

32. C. *purpuraſcens.* The ſpecimen No. 41 in the Dillenian Herbarium, does not appear to me to be diſtinct from C. *roſea*, and I have little doubt that this is the ſpecies here deſcribed by Hudſon. It however ſeems better, eſpecially as the matter is in ſome degree queſtionable, that it ſhould be continued with the name of C. *roſea*, by which it is now univerſally known.

33. C. *noduloſa.* On the authority of Mr. Turner is C. *diaphana.* T. 38.

34. C. *pellucida.* T. 90.

35. C. *vagabunda.* There can, I apprehend, be no doubt, from Dillenius'

deſcription, ſpecimen, and figure, that C. *fracta*, T. 14, is the ſpecies here intended.

36. C. *rupeſtris*. T. 23. The ſpecimen No. 28 of the Dillenian Herbarium, to which Hudſon refers as his variety β of this ſpecies, is much injured, but I have little doubt that it is C. *diffuſa*. T. 21.

37. C. *ſericea*. In the Dillenian Herbarium there are two ſpecimens under the name of C. *marina trichoides virgata ſericea*, of which one is marked "ex aquis dulcioribus," the other "e maritimis;" and of theſe the former is a trifling variety of C. *glomerata*. I have never ſeen C. *glomerata* with a ſimilar appearance to that of the latter, or having the branches ſo much elongated, but from the look of the joints, diſpoſition of the ramuli, and place in which it was gathered, it may probably be Dr. Roth's variety β. *marina* of that ſpecies. According to the *Flora Anglica*, *C. ſericea* grows "in rupibus et ſaxis ſubmarinis;" and I cannot help ſuſpecting that Hudſon confounded C. *læte vireus* with other plants under this name, but neither that ſpecies, nor either of Dillenius's ſpecimens have the leaſt affinity with C. *littoralis*, to which in the *Hiſtoria Muſcorum* it is ſaid to be cloſely allied, and I therefore doubt whether *C. ſericea* can be regarded as a ſingle ſpecies.

38. C. *glomerata*. T. 13.

39. *C. fulva*. I ſuſpect that C. *repens*, T. 18, is the plant here deſigned, but proof is wanting.

40. C. *nigreſcens*. Following the generally received opinion among Botaniſts, I have in my ſynopſis agreed with Dr. Smith in retaining the appellation of C. *nigreſcens* to the plant ſo called in *Engliſh Botany*, though, in ſo doing, I have acted in oppoſition to my own private opinion, and to the authority of Sir Thomas Frankland, who communicated to me a ſpecimen of C. *urceolota* under that name.

41. *C. fuſca*. T. 95.

42. *C. fucoides*. T. 75.

43. C. *villoſa*. T. 37.

44. C. *imbricata*, as above mentioned, is C. *equisetifolia*. T. 54.
45. C. *coccinea*. T. 36.
46. C. *pennata*. T. 86.
47. C. *parasitica* of English Botany, t. 1429, and of my Synopsis.
48. C. *ægagrophila*. T. 87.

---

I AM proud to acknowledge the flattering manner in which most of those Botanists who are distinguished by their knowledge of the submersed Algæ have assisted me in this work. Mr. Turner in the most friendly manner has exerted himself to procure and give me all the information in his power, and to him I am indebted for the descriptions of C. *arbuscula*, *ægagrophila*, *orthotrichi*, and *pellucida*. Sir Thomas Frankland, Bart. and the Rev. Hugh Davies, have obligingly communicated some authentic specimens in their possession, and thereby enabled me to fix the synonymy of several of Hudson's species, with greater certainty than would otherwise have been possible. To James Brodie, Esq. Joseph Woods, junr. Esq. William Jackson Hooker, Esq. William Borrer, junr. Esq. Miss Hutchins, and Mr. William Weston Young, I am indepted for the discovery of many new species, and I am still further indebted to Mr. Hooker and Mr. Woods for several beautiful drawings with which they have favored me, nor must I omit to acknowledge the service which that part of my undertaking has received from the professional talents of Mr. Young. I have also to thank Dr. Turton for his readiness at all times to assist me. To the Right Hon. Sir Joseph Banks I am under great obligation for the free access which he has allowed me to his invaluable Library and Herbarium; and to Dr. Williams, Professor of Botany, at Oxford, for the opportunity he has liberally afforded me of examining the specimens in the Dillenian Herbarium.

# SYNOPSIS OF THE BRITISH CONFERVÆ.

## With Notes, and a Description of the Species not elsewhere mentioned in this Work.

*The descriptions which I have marked with an asterisk are taken from dried specimens.*

A. *Subarticulatæ.* †

1. *dichotoma.* C. filis subarticulatis dichotomis, fasciculatis, strictis, fastigiatis, viridibus; ramis elongatis, remotis. T. 15.

What I have described as capsules under this species, Dr. Roth supposes to be the eggs of insects, and I regret that I have since had no opportunity of re-examining them. Mr. Turner has observed, that when kept but a short time in water they fall off in great numbers, but he says that their appearance is precisely similar to that of the capsules of other *Vaucheriæ.*

† The four first species of this division belong to the Vaucherian genus *Ectosperma*, lately taken up (most injudiciously in my opinion) in English Botany by the name of *Vaucheria.* The able author of the *Hist. des Conf. d'eau douce* has described many plants as distinct species of this genus, of which by far the larger part have been found in Britain, but, as has already been observed in the introduction to this work, p. 17, I have every reason to believe that these, instead of being ranked as species, do not even deserve to be considered as varieties, all of them depending upon the capsules, of which the number and situation vary in the same individual. I have therefore not only here omitted to notice them, as I thought that the so doing would unnecessarily swell the number of my species, but I even doubt whether of the four here described the three latter are specifically distinct from each other.

2. *amphibia.* C. filis ſubarticulatis, ramoſis, denſiſſimè implexis, obſcurè viridibus; ramis ſparſis, patentibus, remotis. T. 43.

3. *veſicata.* C. filis ſubarticulatis, ramoſis, rigidis, fuſco-viridibus; veſiculis innatis, ſolitariis, ellipticis, filamento latioribus. T. 74.

   *Ectoſperma ſeſſilis.* VAUCHER. *Hiſt. des Conf.* p. 31. t. 2. f. 7. *Eng. Bot.* t. 1765.

4. *Dillwynii.* C. filis ſubarticulatis, procumbentibus, ramoſis, viridibus; ramis ſubdichotomis, alternis.

   *C. frigida.* T. 16.

   C. *Dillwynii.* WEBER and MOHR. *Groſs. Brit. Conf.* II. p. 14. t. 16.

   *Ceramium Dillwynii.* ROTH. *Cat. Bot.* III. p. 117.

   *Ectoſperma terreſtris.* VAUCHER. *Hiſt. des Conf.* p. 27. t. 2. f. 3.

   C. *frigida* of Roth, to which I had erroneouſly referred this ſpecies, is my C. *muralis*, and probably the plant deſigned in Engliſh Botany under the name of *V. geminata*, t. 1766, is nothing more than this ſpecies, as I have ſeen ſimilar fruit both upon *C. amphibia* and C. *Dillwynii.*

5. *Myochrous.* C. filis ſubarticulatis, ramoſis, implexis, fuſcis; ramis ſimplicibus, ſubſecundis, geminis, incurvis. T. 19.

   Since the publication of this ſpecies, Mr. Woods and myſelf have found it in various parts of Wales, and Miſs Hutchins has gathered it in the neighbourhood of Bantry. Mr. Hooker and Mr. Borrer brought laſt year from the cave in the Iſle of Skye, called Sloch Altramine, a variety of this ſpecies, of a dark green color and looſe mode of growth, with filaments above an inch long.

6. *Comoides.* C. filis ſubarticulatis, ramoſis, ferrugineis; ramis ſparſis, remotiuſculis, apice acuminatis. T. 27.

   Since I publiſhed the deſcription of this ſpecies, it has been found at Yarmouth by Mr. Turner, and in Suſſex by Mr. Borrer, and by Mr. Woods at Dover. It ſeems probable that C. *rufa* of Roth's *Cat. Bot.* III. p. 280, is the ſame plant, in confirmation of which, and of the opinion given in the note at the beginning of this fection, and in the introduction as to the *Ectofperma*, I copy the following Extract from a Letter from the late Dr. Mohr to Mr. Turner.

" Taking this plant for the true C. *comoides* of Dillwyn, I foaked my original fpecimen of it, and what did I find? An *Ectofperma* of the Rev. M. Vaucher, but as Dr. Roth has remarked, without capfules as they are called. I hardly think there is more than one fpecies of *Ectofperma* in the world, (which may feem very paradoxical) but if there are more to be diftinguifhed, you will allow it can only be done by regarding the Vaucherian grains or Rothian capfules."

B. *articulatæ, filis cylindricis.*

a. *articulis breviffimis.*†

* *fimplices.*

7. *fontinalis.* C. filis fimplicibus ftrictis, brevibus fafciatis, atro virentibus; diffepimentis diftinctis; articulis breviffimis. T. 64.

From the defcriptions in the third fafciculus of the *Catalecta Botanica*, it may be doubted whether Dr. Roth's C. *limofa* is not C. *fontinalis* of Hudfon, and vice vêrfa.

8. *limofa.* C. filis fimplicibus, ftrictis, tenuiffimis, fafciatis lubricis, mucofis, cœruleo-virentibus; diffepimentis obfoletis; articulis breviffimis. T. 20.

9. *decorticans.* C. filis fimplicibus, curvis, tenuiffimis fafciatis, denfiffimè contextis, cœruleo-virentibus; diffepimentis obfoletis; articulis breviffimis. T. 26 and T. *A*.

C. *violacea.* *Fl. Ang.* p. 592.

C. *confragofa.* *Fl. Scot.* p. 976.

† In this divifion are comprehended the *Oscillatoria* of Vaucher, a moft diftinct and natural tribe of Confervæ, which will in all probability hereafter form a feparate genus. I exceedingly regret that I have not been able to find a more happy definition of this divifion, not knowing, as is already obferved in the Introduction, how to characterife the particular ftructure of the joints, which feem unlike thofe of all other Confervæ.

*C. mucosa confragosis rivulis innascens.* DILL. Hist. Musc. p. 15. t. 2. f. 4.

In the supplementary plate *A* I have given a highly magnified drawing by Mr. Young, of the appearance of capsules on this species, which is described in my introduction. Although there can be scarcely any doubt of the propriety of the above references, as has been mentioned in my remarks on Hudson's species, yet more than one species having been described under the name of *violacea*, I have thought it best to retain that of *decorticans*, by which it is now generally known, and which is very characteristic of the plant.

10. *muralis.* C. filis simplicibus, curvis, longis, rigidiusculis, sparsis, fasciatis viridibus; articulis brevissimis. T. 7.

C. *muralis.* ROTH. *Cat. Bot.* III. p. 189.

*C. frigida.* *Cat. Bot.* I. p. 166. *Fl. Germ.* III. pars. 1. p. 491.

11. *confervicola.* C. filis simplicibus, abbreviatis, fasciculatis, liberis, fasciatis, intensè æruginosis, apice acuminatis; articulis brevissimis. T. 8. and T. *A*.

Since the publication of this species, Mr. Hooker has discovered on some specimens, capsules surrounded by a pelludid limbus, and transverely divided by a pellucid line in the same manner as those of C. *interrupta*. The acuminated apices of this species and of C. *scopulorum*, have always made me doubtful whether they should be regarded as true *Oscillatoriæ*, and this suspicion has been strengthened by Mr. Hooker's discovery. For the highly magnified sketch of one of these capsules, made from a recent specimen and given in my supplementary plate *A*, I am indebted to Mr. Hooker.

12. *scopulorum.* C. filis simplicibus, curvis, abbreviatis, fasciatis, atro-virentibus, bali per viscositatem cœherentibus, apice attenuatis; articulis brevissimis. T. *A*.

*C. scopulorum.* WEBER and MOHR *Reise durch Schweden*, p. 195. T. 3. f. 3. ROTH, *Cat. Bot.* III. p. 191.

On Planks in the Sea, between Bognor and Aldwick; *Mr. Borrer.*

Rocks by the Sea fide at Cawfie, Murrayfhire; *Mr. Hooker* and *Mr. Borrer.*

Mr. Hooker by comparing the plants gathered in the above mentioned places with Mr. Turner's authentic fpecimens from Dr. Mohr, afcertained the propriety of the prefent reference. It is nearly allied to *C. confervicola*, but differs in its far darker color, fhorter filaments, and in the fingular manner by which they appear agglutinated together towards the bafe. The drawing in the fupplementary plate A was made from a dried fpecimen by Mr. Hooker. The plant is reprefented of its natural fize, and alfo when magnified with powers 3 and 2 of a compound microfcope.

** *coadunatæ.*

13. *vaginata.* C. filis in vaginâ ramofo-fafciculatis, abbreviatis, cœruleo-virefcentibus; articulis breviffimis. T. 99.

*C. vaginata.* Eng. Bot. t. 1995.

Since I publifhed the defcription of this fpecies, it has been found by Mifs Hutchins, growing on *Hypnum prælongum* in the neighbourhood of Bantry.

14. *mirabilis.* C. filis fpurie ramofis, breviufculis ftrictis, cœruleo-virefcentibus; ramis e coadunatis genuflexuris filamentorum; articulis breviffimis. T. 96.

15.* *majufcula.* C. filis fpurie ramofis, crifpatis, elongatis, laxè implicatis, atro virefcentibus; ramis e filamentis coadunatis; articulis breviffimis. T. *A.*

In the Sea. On Santon Sands, near Plymouth; *Mifs Hill.* Bantry Bay; *Mifs Hutchins.*

This fpecies is nearly allied to *C. diftorta* and *C. mirabilis*, the branches being fometimes united in the manner of the former, and fometimes as in the latter. It may be diftinguifhed from both of thefe, as well as from the other *Ofcillatoriæ*, by its remarkably curled and twifted filaments, and by their ſomewhat greater diameter. It grows in thick tufts, not unfrequently three inches in length, and of a very dark blackiſh green color. For the drawing, which is made from a dried ſpecimen, and repreſents the filaments when magnified with powers 2 and 1, I am indebted to my friend Mr. Hooker.

16. *diſtorta.* C. filis ſpurie ramoſis, ſub-ſtrictis, cœruleo-vireſcentibus; ramis e filamentis coadunatis, diſtortis; articulis breviſſimis. T. 22 and T. *A.*

The figure of this plant, T. 22, is erroneous as far as relates to the branches, which inſtead of being as there repreſented, appear rather to be merely different filaments united together in the ſame way as thoſe of C. *mirabilis.* The affinity between theſe two ſpecies is very ſtrong, and the leading difference ſeems to be that in C. *mirabilis* the ſides of the two filaments are joined and continue longitudinally united, whereas in C. *diſtorta* the end only of one filament is attached to the ſide of another. This curious union is repreſented in my ſupplementary plate A, as it appears with power 1 of the microſcope, and alſo the fructification which I have deſcribed in the introduction to this work.

b. *articulis longis.*

* *ſimplices.*

17. *zonata.* C. filis ſimplicibus, tenuibus, lubricis, vireſcentibus; articulis diametrum longitudine vix ſuperantibus, granulis in faſcias latas coarcervatis

C. *zonata.* WEBER and MOHR. *Reiſe durch Schweden*, p. 97. T. 1. f. 7. a. b. ROTH. *Cat. Bot.* III. p. 269.

C. *lubrica.* T. 47.

Found lately at Lound, near Yarmouth, by Mr. Hooker, and in Suſſex by Mr. Borrer.

18. *rivularis.* C. filis ſimplicibus, tenuibus, longiſſimis, densè compactis, plerumque contortis, intensè viridibus; articulis diametro ſeſqui longioribus. T. 39.

β. *aculeata.* Spinulis ramuliformibus. T. *A.*

I first discovered the present supposed variety in company with my friend Joseph Woods, junr. in some dark shady rills on Finchly Common, and afterwards in a shady well on Stamford Hill, and in a similar well near Yarmouth. It may be at once distinguished by the naked eye from the more common state of *C. rivularis* by its still darker color, but under the microscope it appears to differ only in its numerous short spine-like processes, of which the joints resemble those of the main filament, except that they become gradually narrower, and at length terminate in a fine point. These thorn-like processes bear a considerable resemblance to the ramuli of *C. lubrica*, both in the size, shape and irregularity of their disposition, but of their nature I am still unable to satisfy myself further than that for reasons given in my introduction, they are not occasioned by a proliferous germination. The drawing at figure 3 of the supplementary plate A, was made in 1802, with power 1 of my microscope, from the plant which I then gathered near Finchly. With the sketches marked 1 and 2 (of which the former represents the plant when slightly, and the latter when highly magnified) I have been favored by Mr. Woods, who has since discovered this appearance of the species in several places about London. It grows not like the foregoing in springs, but in pools and ditches which are dried up early in the summer, and ought perhaps to be regarded as a separate variety. Some of the filaments are entirely simple, and these resemble those of C. *rivularis;* in others there are a few acuminated processes similar to those above mentioned, whilst others are beset with crowded processes of various lengths, and of these the longest are less acuminated than the others, and are again sometimes furnished with other extremely short secondary spines.

19. *bipartita.* C. filis simplicibus, tenuibus, longissimis densè compactis, flavo-virentibus; articulis diametro sub-triplo longioribus, demum bipartitis. T. 105.

20. *fugacissima.* C. filis simplicibus tenuibus flavo-virentibus; articulis pellucidis medio sæpe granulis fasciatis, diametro sub-sesquilongioribus. P. *B.*

*C. fugacissima.* ROTH. *Cat. Bot.* III. p. 176.

Frequent in Pools and Ditches, adhering to glass and other substances.

In Mr. Turner's Herbarium there are two specimens from Dr. Roth, marked *C. fugacissima*, of which one belongs to *C. sordida*, and the other to the present species. It is most nearly allied to *C. sordida*, but may be at once distinguished by its far shorter joints. By drying *C. fagacissima* loses its color, and gradually becomes of a dirty white. The sketch at Plate B. represents a filament magnified 1.

21. *sordida.* C. filis simplicibus, tenuibus, flavo virentibus; articulis pellucidis, diametro quadruplo longioribus. T. 60.

22. *alternata.* C. filis simplicibus, tenuibus, glauco virescentibus; articulis hic illic inflatis, alternatim pellucidis obscurisque, diametro sesquilongioribus.

*Prolifera vesicata.* VAUCHER. *Hist. des Conferves*, p. 132. t. 14. f. 4. (exc. syn.)

β. *fuscescens*, filis fuscescentibus, T. *B.*

In a rivulet near Swansea; β. In ditches at Stoke Newington; *Mr. Woods.* On decayed leaves in the ditches at Heigham, near Norwich; *Mr. Hooker.* Ditches about Belfast; *Mr. Templeton.* Pools near Bantry; *Miss Hutchins.*

The filaments grow in loosely entangled masses, six or eight inches in length, and are of about the same diameter as those of *C. sordida.* The color of the plant, which I once gathered near Swansea, agreed with Vaucher's description and was of a glaucous green. The joints are alternately opake and pellucid, and some of them in almost every filament are remarkably inflated, by which this species may be readily distinguished from its congeners. The variety β. appears to differ in no other respect than in being of a brown color, and of this Mr. Hooker favored me with the magnified ſketch given in my ſupplementary plate B.

23. *faſciata.* C. filis ſimplicibus, tenuibus, mucoſis, purpureo-fuſcis; articulis medio faſciâ anguſtâ tranſverſim notatis, longitudine diametrum æquantibus. T. *B.*

On decaying ſticks, leaves, &c. in a ditch at Stoke Newington; *Joſeph Woods, junr. Eſq.*

Mr. Woods diſcovered this ſpecies growing in ſlippery maſſes about one and a half inch long, of a purple brown color, and forming a thick coat over decaying ſubſtances in a ditch at Stoke Newington. The length and diameter of the joints is equal, and in the middle of each there is the appearance of a dark narrow tranſverſe band, which however proceeds from the internal organization of the plant, and therefore appears ſomewhat ſhorter than the diameter of the filament. For the drawing in my ſupplementary plate B, which was made with power 1 of the microſcope, I am indebted to my friend Mr. Woods.

24. *lineata.* C. filis ſimplicibus tenuibus, fragilibus, fuſcis; diſſepimentis contractis; articulis lineâ unâ alterave tenuiſſimâ tranſverſim ſtriatis, diametro ſub-triplo longiorbus. T. *B.*

Among the leaves of water plants in the River Lea at Walthamſtow.

In March, 1802, I found a ſingle ſmall ſpecimen of this ſpecies among a jelly-like ſubſtance of the Tremella kind, which almoſt covers the water plants in the Lea at Walthamſtow. The filaments are ſimple, very brittle, contracted at the diſſepiments, and of a brown color. The length of the joints in ſome filaments is about thrice, and in others not more than twice the diameter, and they are generally marked with one or two tranſverſe lines at uncertain diſtances from each other. I have not ſince been able to find more than a few imperfect filaments of this plant, and in one of theſe now before me, I obſerve one or two joints much ſhorter than the others, whoſe length ſcarcely exceeds the diameter, and which in appearance ſomewhat approach thofe of the following fpecies before they affume their oval form. The general appearance of the two plants is however entirely diffimilar, but Dr. Roth's account of the wonderful changes which he has obferved in his C. *annulina*, almoft induce me to fufpect that they may poffibly both belong to the fame fpecies. For the lower of the two fketches in Tab. B, I am indebted to my friend Mr. Woods, and they are both made with the higheft power of a compound microfcope.

25. *nummuloides*. C. filis fimplicibus, tenuibus, fragilibus, fufco aureis; articulis diametro fub-brevioribus, demum in glomerules fub-ovales, moniliformes, approximates mutatis. T. *B.*

Among the leaves of water plants in the River Lea at Walthamftow.

In March, 1802, I found a few detached filaments of the prefent plant, mixed with thofe of C. *lineata*; among the Tremella-like flime with which, as before mentioned, many of the plants in the River Lea are covered. I have not difcovered any filaments which appear to be at all perfect, but they feem fufficiently fo to prove that the plant differs materially from every other Britifh fpecies, and by publifhing this imperfect account I truft that I fhall induce fome other Botanift to fearch for it, and more completely afcertain its nature. The filaments are cylindrical, of a brittle nature, and reddifh, yellowifh, or yellowifh brown color. The internal veficles which conftitute the joints appear to be at firft cylindrical, but at length collapfe into an oval form, fo as to give the filaments when highly magnified, fome refemblance to a feries of guineas. The length of their joints is generally fomewhat lefs than their diameter. C. *nummuloides*, although fpecifically diftinct, appears to poffefs fome affinity with a fpecies figured in the 4th Vol. of the Stockholm Tranfactions, under the name of *C. moniliformis*. The drawing in the fupplementary plate B, reprefents the filaments when magnified with power 1.

26. *punctalis*. C. filis fimplicibus, tenuiffimis, longis, viridefcentibus; articulis diametro fub-duplo longioribus, fucco in globulum folitarium demum congefto. T. 51.

27.* *Mucosa.* C. filis simplicibus, tenuissimis, lubricis, luteo virescentibus; articulis sub-torosis, longitudine diametrum æquantibus. T. *B.*

In stagnant Pools about Bantry. *Miss Hutchins.*

The gelatinous nature of this Confervæ makes it very difficult to investigate its real nature after it has been dried, in which state alone I have at present seen it. It is then scarcely distinguishable by the naked eye from *C. spiralis*, which it resembles both in its color, the mode of its growth, and the size of its filaments, though under a microscope the internal structure appears so widely dissimilar. There is however a strong peculiarity in it even in this state, that its excessively gelatinous texture prevents the filaments from cohering together, or even touching each other, and they lie quite distinct upon the paper. Miss Hutchins remarks that it has when recent a beautiful color. For the magnified drawing in my supplementary plate B, I am indebted to Mr. Hooker, but it was unavoidably made from a specimen which had been dried.

28.* *implexa.* C. filis simplicibus, crispato-implexis, tenuibus, mollibus, intensè lurido viridibus; articulis diametro sesquilongioribus. T. *B.*

On Rocks in the Sea near Bantry. *Miss Hutchins.*

This species is nearly allied to *C. tortuosa*, but the filaments are more entangled and slender, the texture less rigid, and the joints shorter. The drawing in plate B was made by Mr. Hooker, with power 1 of his microscope, from a specimen which had been dried.

29. *tortuosa.* C. filis simplicibus, rigidiusculis, crispatis, implicatis tenuibus intensè viridibus; articulis diametro driplo longioribus. T. 46.

30. *crispa.* C. filis simplicibus, rigidiusculis, crispatis, proliferis, laxè implicatis, crassiusculis, viridibus; articulis diametrô sub-triplo longioribus, siccitate alternatim compressis. T. *B.*

*C. capillaris.* *Sp. Plant.* p. 1636 (excl. Syn.) ROTH. *Fl. Germ.* III. pars. 1. p. 502. *Cat. Bot.* III. p. 261.

*Prolifera crispa.* VAUCHER. *Hist. des Conf. d'eau douce*, p. 130. t. 14. f. 2.

In a rapid ſtreamlet at Coſteſy, Norfolk. *W. J. Hooker, Eſq.*

Mr. Hooker, who alone has diſcovered this ſpecies in Britain, informs me that he has ſeen the filaments carried out by the current to the length of fifteen or twenty feet: their thickneſs is ſomewhat greater than that of C. *tortuoſa*, from which it may be at once diſtinguiſhed by its longer joints, as well as by the curious manner in which they become alternately compreſſed when the plant is dried without preſſure. Mr. Hooker has diſcovered lateral acuminated proceſſes iſſuing from the filaments, preciſely ſimilar to thoſe which Vaucher has figured on his *Prolifera criſpa*, and the plant in other reſpects ſo far accords with his deſcription, as to leave no doubt of the propriety of the above reference. He informs me that there is a ſpecimen of this ſpecies preſerved in the Linnæan Herbarium, with the name of C. *capillaris*, but Linnæus in his deſcription refers to the *Hiſtoria Muſcorum*, and it is certain that the ſpecies there figured is what I have repreſented at T. 9. I cannot therefore ſee the neceſſity for any alteration, which as that plant is now almoſt univerſally known by the name of *capillaris* would in my opinion only tend to confuſion. The drawing at plate B, for which I am indebted to the liberality of Mr. Hooker, repreſents C. *criſpa* magnified 2.

31. *capillaris.* C. filis ſimplicibus, rigidiuſculis, criſpatis, fragilibus, laxè implicatis, craſſis, viridibus; articulis diametrum longitudine vix æquantibus. T. 9.

*β. minor.* Fills triplo tenuioribus.

For reaſons given in the foregoing obſervations on C. *criſpa*, my former reference to the ſpecies Plantarum ſhould have been omitted. Mr. Hooker favored me with ſpecimens of what I have here arranged as a variety, which he diſcovered growing mixed with C. *criſpa*, far from the neighbourhood of the ſea, in the river at Helleſdon, near Norwich. The filaments are thrice more ſlender than thoſe of C. *capillaris*, which with its different place of growth, ſeems to indicate that it ſhould conſtitute a ſeparate ſpecies, and I regret therefore that I am unable to diſcover any other diſtinctive mark whatſoever.

32. *ærea.* C. filis ſimplicibus, rigidis, ſtrictis, craſſis, praſinis; articulis diametro brevioribus, demum bipartitis. T. 80.

*β. lubrica.* Filis lubricis, mollibus.

C. *ærea.* *Eng. Bot.* t. 1929.

This curious variety, which was found on the Yarmouth Beach by Mr. Hooker, in the ſpring of 1808, attached to a piece of deal, differs ſo extraordinarily from the common appearance of C. *ærea*, that except under a microſcope nobody would ſuſpect them of being the ſame. It grew in a very large tuft, and its filaments were remarkably ſoft, tender, ſlippery and gloſſy, ſo as to float with the ſlighteſt agitation of the water and adhere cloſely to paper and glaſs in drying.

33.* *Melagonium.* C. filis ſimplicibus, rigidis, ſtrictis, craſſis, praſinis; articulis diametro ſub-triplo longioribus. T. *B.*

C. *Melagonium.* Weber and Mohr. *Reiſe durch Schweden*, p. 194, t. 3. f. 2. a. 6. Roth. *Cat. Bot.* III. p. 254.

In the Sea, near Newton Nottage, Glamorgan; *Mr. Young*. Near Bantry, not common; *Miſs Hutchins*. Once found on the ſhore near Swanſea.

The mode of growth, color and habit of this plant, which was firſt diſcovered on the coaſt of Sweden by Meſſrs. Weber and Mohr, are preciſely ſimilar to thoſe of C. *ærea*, from which it differs in the ſomewhat greater thickneſs of its filaments, and greater length of its joints. It was C. *melagonium* of which ſome years ago I found a ſingle filament on the ſhore near Swanſea, and which I then conſidered as a variety of C. *ærea*, and as ſuch it is mentioned in my deſcription of that ſpecies. The drawing at T. B. was made by Mr. Hooker from a dried ſpecimen, and repreſents the plant of its natural ſize, and alſo when magnified 3.

** *conjugatæ.* †

34. *nitida.* C. filis ſimplicibus, demum conjugatis, atro viridibus, ſplendenter lubricis; granulis in ſpiras plures, arctas, diſpoſitis; articulis diametrum longitudine ſub-æquantibus. T. 4. f. C.

C. *nitida.* *Fl. Dan.* t. 819, and *C. jugalis,* t. 883.

*Conjugata princeps.* VAUCHER. *Hiſt. des Conf. d'eau douce,* p. 64. t. 4.

35. *decimina.* C. filis ſimplicibus, demum conjugatis viridibus, ſplendenter lubricis; granulis in ſpiras duas laxas diſpoſitis; articulis diametro ſexduplo longioribus.

C. *decimina.* MULLER in *Nova Acta Petrop.* III. p. 94. t. 2.

C. *nitida.* T. 4. f. A. B. and *C. jugalis,* T. 5.

C. *ſetiformis β.* ROTH. *Cat. Bot.* and *Fl. Germ.*

It will be immediately perceived that the ſpecific characters which ſeparate this ſpecies from the foregoing, lie in the different lengths of their joints, the very diſſimilar arrangement of their ſpires, and the dark almoſt black green of the one contraſted with the paler hue of the other. In both theſe ſpecies, Mr. Turner and myſelf have obſerved that the granules are ſometimes found, either from peculiarity of ſituation or from diſeaſe, ſcattered irregularly all over the joints, inſtead of preſerving their natural ſpiral diſpoſition; and in ſome individuals there are no traces of theſe whatever, though at the ſame time there is no appearance of their ever having been conjugated.

36. *longata.* C. filis ſimplicibus, demum conjugatis, flavo-virentibus, lubricis; granulis in ſpiram unicam laxam diſpoſitis; articulis diametro quadruplo longioribus.

† In this diviſion are compriſed the Confervæ referred by Vaucher to his genus *Conjugata,* a particularly natural and intereſting family, which I have deſcribed in my Introduction, p. 17. I am ſorry that I cannot follow this excellent Botaniſt in adopting all the ſpecies which he has deſcribed, but I have been led by my own obſervations to divide *C. nitida, C. ſpiralis,* and *C. bipunctata* each into two ſeparate ſpecies, in doing which I hope I have been correct, though I am far from feeling certain on the ſubject.

*C. longata.* VAUCHER. *Hiſt. des Conf. d'eau douce*, p. 71. t. 6. f. 1.

A part of the filaments repreſented in T. 3. f. A. belongs to this, and a part to the following ſpecies. The ſpires are ſometimes though rarely, double, but even in this ſtate it may be diſtinguiſhed from C. *decimina* by its more ſlender filaments and ſomewhat ſhorter joints. I have never ſeen a ſpecimen of *C. inflata* of Engliſh Botany, but am led by the deſcription and figure to ſuſpect that it is not diſtinct from this ſpecies.

37. *ſpiralis.* C. filis ſimplicibus, demum conjugatis, flavo-virentibus; granulis in ſpiram unicam compactam diſpoſitis; articulis diametro ſub-duplo longioribus. T. 3. f. C. and T. C.

C. *porticalis.* VAUCHER. *Hiſt. des Conf.* p. 66. t. 5. f. 1. (exc. Syn. Mulleri.)

Since the Introduction was printed, a curious ſpecimen of this ſpecies has been gathered by Mr. E. Horne, at Clapham, and examined by Mr. Woods, who gives the following account of it. "The plant is a pale dirty green nearly without gloſs, about the uſual ſize of *C. ſpiralis;* when magnified, the length of the joints is ſeen to be about equal to their width or a little more, and the ſpiral tube is in moſt parts nearly obliterated, but the chief ſingularity of this plant is in the connecting proceſſes which are uniformly at the ends, inſtead of as uſual in the middle of the joints; and each of which appears to unite with the proceſs of the next joint of the ſame filament. No indication of the conjugation of two filaments is to be obſerved; the dark globules appear only where the two joints are thus connected, and the adjacent one is uniformly empty.

38. *bipunctata.* C. filis ſimplicibus, demum conjugatis, viridi flaveſcentibus, lubricis; articulis bipunctatis, diametro ſub-ſeſqui longioribus. T. 2.

VAUCHER.

Mr. Hooker informs me that he has lately found this ſpecies, with the joints ſeparated like thoſe of C. *flocculoſa*, and that, when ſeparated, the joints became rounded at the corners, and the internal maſſes completely fpherical. Soon after my defcription of this fpecies went to prefs, Mr. Woods difcovered it with the filaments conjugated.

39. *decuffata.* C. filis fimplicibus, demum conjugatis, lutefcentibus, lubricis; articulis bipunctatis diametro fub-triplo longioribus.

*C. decuffata.* VAUCHER. *Hift. des Conf.* p. 76. t. 7. f. 3.

This fpecies is found in the fame fituations and is clofely allied with C. *bipunctata*, but may be diftinguifhed by its more flender filaments, the fmaller fize of the fpots, and the greater length of its joints.

40. *genuflexa.* C. filis fimplicibus, demum hic illio genuflexis, conjugatifque, fragilibus, flavefcentibus lubricis; granulis in lineas horizontales coarcervatis. T. 6. and T. *C.*

The feed defcribed in my Introduction, is reprefented in Plate C. magnified 1.

*** *anaftomofantes.*

41. *reticulata.* *C.* filis anaftomofantibus, reticulatis, in maculas fub pentagonas coadunatis. T. 97.

*C. articulis folutis.*

42. *diffiliens.* C. filis fimplicibus, ftrictis, fragilibus, lœtè viridibus; diffepimentis plerumque folutis; articulis diametro dimidio brevifiribus. T. 63.

43. *pectinalis.* C. filis fimplicibus, ftrictis fragilibus, compreffis cinereis, plerumque acuminatis; diffepimentis fœpe folutis; articulis diametro triplo brevioribus, medio pellucidis. T. 24.

Drs. Mohr and Weber, in their German tranflations of this work, exprefs their opinion very decidedly in favor of uniting *C. pectinalis* and *C. flocculofa*, but I muft confefs I have feen nothing to induce me to depart from my former fentiments that they are quite diftinct.

This and the following fpecies of the fame divifion belong to the genus *Diatoma* of Decandolle, and are by means of *C. diffiliens* united to the other Confervæ.

44.* *teniæformis.* C. filis fimplicibus, compreffis, dilutè viridibus; diffepimentis folutis; articulis diametro dimidio brevioribus, obfoletè variegatis, demum refractis.

*C. teniæformis. Eng. Bot.* t. 1883.

On *Conferva fucoides* in the Sea at Beachy Head. *Mr. Borrer.*

45.* *ftriatula.* C. filis fimplicibus, compreffis dilutè viridibus; diffepimentis alternatim folutis; articulis diametro vix brevioribus, tranfverfim ftriatis.

*C. ftriatula.* Eng. Bot. t. 1928.

On Fuci and Confervæ in the Sea at Cromer; *Mr. Hooker.* At Brighton, *Mr. Borrer.*

46.* *Biddulphiana. C.* filis fimplicibus, compreffis, longitudinaliter ftriatis, viridibus; diffepimentis folutis; articulis quadratis, tranfverfim fafciatis, fub-alternatim refractis.

C. *Biddulphiana.* Eng. Bot. t. 1762.

On Marine Algæ at Southampton. *Mifs Biddulph.*

This plant, which as well as the two former and *C. obliquata*, is here introduced upon the authority of Englifh Botany, appears to be as Dr. Smith obferves, really an extraordinary production, but it feems fcarcely poffible that all the figures in that plate fhould belong to the fame plant, or if they do, does it not lead to a fufpicion that the fpecies of this family have been unneceffarily multiplied by authors?

47. *flocculofa. C.* filis fub-fimplicibus, compreffis, fafcia longitudinali percurfis, cinereis; diffepimentis folutis; articulis quadratis, tranfverfim ftriatis, alternatim refractis. T. 28.

48.* *obliquata.* C. filis ramofis, compreffis, flexuofis, fufco albidis; diffepimentis folutis; articulis quadratis, obliquis, tranfverfim fafciatis, maculatis, alternatim refractis.

*C. obliquata. Eng. Bot.* t. 1889.

On Fuci and Confervæ in the Sea. *Mifs Biddulph.*

*C. articulatæ. filis ſetaceis.*

a. *aveniæ.*

* *ſimplices.*

49. *flacca.* C. filis ſimplicibus, tenuibus, flaccidis, lœti viridibus; diſſepimentis pellucidis; articulis diametro pâullo brevioribus. T. 49.

C. *penicilliformis.* ROTH. *Cat. Bot.* III. p. 271?

50. *Youngana.* C. filis ſimplicibus, cœſpitoſis, flaccidis, obtuſis, lœtè viridibus; articulis utrinque contractis longitudine diametrum æquantibus. T. 102.

*C. iſogona. Eng. Bot.* t. 1930.

51. *curta.* C. filis ſimplicibus faſciculatis, ſub-cartilagineis, abbreviatis, utrinque alternatis, fuſco-olivaceis; diſſepimentis pellucidis; articulis diametro ſub-longioribus. T. 76.

52.* *flaccida.* C. filis ſimplicibus, fâſciculatis, abbreviatis, flaccidis, baſi latioribus apicem verſus attenuatis olivaceo viridibus; articulis inferioribus diametro dimidio brevioribus, ultimis æquantibus. T. C.

On Fucus fibroſus on Santon Sands, Devon. *Miſs Hill.*

This ſpecies appears to have been gathered only by Miſs Hill, who communicated it to Mr. Turner. It grows in ſmall tufts about half an inch long, and may be diſtinguiſhed from *C. curta* by its flaccid nature, and from *C. fucicola*, as well as all its other congeners, by the rather abrupt manner in which the joints of the upper part of the filament increaſe in length to double that of the lower part. Its ſubſtance is ſomewhat gelatinous, and in drying it adheres, though not very firmly, to either Glaſs or Paper. In the drawing at Plate *C*, for which I am indebted to Mr. Hooker, the plant is repreſented of the natural ſize, and when magnified with power 3, the upper and lower part of the filament are alſo ſeparately repreſented, magnified 2.

53. *fucicola.* C. filis ſimplicibus, faſciculatis, breviuſculis, obtuſis, ferrugineis; diſſepimentis pellucidis; articulis diametro duplo-longioribus. T. 66.

54. *carnea.* C. filis ſimplicibus, tenuibus, abbreviatis, carneis; articulis toroſis, diametro ſub-triplo longioribus; ſucco in globulum ſolitarium congeſto. T. 84.

55. *ericetorum.* C. filis ſimplicibus, procumbentibus, implexis, fuſco-violaceis; articulis diametro duplo longioribus, demum ſub-ovalibus. T. 1.

Dr. Roth in the laſt volume of his *Catalecta Botanica*, has deſcribed this ſpecies as branched, but I have never ſeen it ſo.

56. *fuſco-purpurea.* C. filis ſimplicibus, ætate inæqualiter toroſis, fuſco-purpureis; articulis diametro dimidio brevioribus, demum ſerie globulorum cinctis. T. 92.

57.* *atro-purpurea.* C. filis ſimplicibus, ætate inæqualiter toroſis, atro purpureis; articulis diametro dimidio brevioribus, demum ſerie duplici globulorum cinctis. T. 103.

Since the publication of this plant, ſpecimens have been found on the Coast of Cornwall by Mr. W. Raſhleigh, and communicated by him to Mr. Turner.

** *ramoſæ.*

58. *feneſtralis.* C. filis ramoſiſſimis, repentibus, minutiſſimis, centrifugis, albidis; ramis plerumque divaricatis; diſſepimentis ſub-obſoletis. T. 94.

59.* *nivea.* C. filis ramoſis, tenuiſſimis, rigidiuſculis, niveis; ramis in verticello confertis; articulis diametrum longitudine ſub-æquantibus. T. C.

*Byſſus lanuginoſa.* Willan, *Obſ. on Sulphureous Waters.* p. 10.

In Sulphur Springs. At Croft, Yorkſhire, and Dinſdale, Durham; *Dr. Willan.* At Middleton One Row, near Darlington; *Mr. Backhouſe.*

Although I have not ſeen any other ſpecimens of *C. nivea*, than thoſe which I received from Darlington, yet from Dr. Willan's deſcription there can be no doubt that it is the plant which he has deſcribed. Dr. Willan ſays it is a remarkable circumſtance that this ſpecies is found below the ſpring, no further than the water retains the ſenſible ſulphureous qualities, as if the hepatic gas was neceſſary to its production and nouriſhment.

It grows on roots and other fubftances, which it covers with white filaments two or three lines in length, and fo extremely flender, that under the higheft power of my microfcope, their thicknefs fcarcely appears equal to that of horfe-hair. Some of the filaments are fimple, but moft of them are fingularly befet towards the middle with a whirl-like clufter of very numerous fimple branches refembling proliferous fhoots. Diffepiments with a very high power are clearly difcernable, and they divide the filaments into joints, the length and thicknefs of which are about equal. The drawing at table C, for which I am indebted to my friend Jofeph Woods, reprefents the plant of the natural fize and when magnified 2. A fragment is alfo added (on a rather larger fcale than it appeared with the higheft power) to fhew the joints.

60. *ochracea.* C. filis ramofiffimis, tenuiffimis, perfragilibus, denfiffimè compactis, gelatinam ochraceam tamen in floccos fecedentem conftituentibus, diffepimentis fub-obfoletis. T. 62.

61. *lactea.* C. filis ramofis in maffam informem gelatinofam confertis, hyalinis, fordidé lacteis; ramis e quovis diffepimento; articulis longiffimis. T. 79.

62. *typhloderma.* C. filis fub-ramofis in pelliculam olivaceam denfiffimé implexis; articulis longitudine diametrum æquantibus. T. 83.

63. *fanguinea.* C. filis ramofis in pelliculam gelatinofam fanguineam, denfiffimè implexis; ramis divaricatis; articulis diametro fefquilongioribus.

Mr. Young difcovered the prefent fpecies, forming a denfely matted membrane on the furface of fome Ifinglafs fize, in which he had put a quantity of patent yellow to diffolve, but we have fince repeatedly endeavored to produce it in the fame manner without fuccefs. Its dark crimfon color is of itfelf fufficient to diftinguifh it from its congeners.

64. *pallida.* C. filis dichotomis, curvato-flexuofis, faftigiatis, in pelliculam gelatinofa-coriaceam implexis, pallidè ochraceis; dichotomarum angulis rotundatis; articulis longiffimis. T. 78.

65.* *arachnoidea.* C. filis ramofis, tenuibus, in membranam arachnoideam laxè implicatis, pallidè flavefcentibus; ramis fparfis, remotis, fimplicibus; articulis longitudine variantibus, diametrum fub-quadruplo fuperantibus. T. C.

On decayed Trees in the Wood at Croftwick near Norwich. *Mr. Hooker.*

I cannot find that this fpecies has been noticed either as a Conferva or Byffus, in which latter genus it would have been moft probably arranged by the older authors. It forms a fine fpider-like web on decaying wood of a light yellow color. The filaments are branched, extremely flender, flaccid, and loofely entangled: the branches are fimple, remote and difpofed without apparent order: the diffepiments are of a dark color, and divide the filaments into joints, whofe length, though variable, is moft ufually about four times greater than the diameter. I am not aware of its having been found by any other Botanift than Mr. Hooker, and to him I am indebted for the drawing of plate C, which reprefents the plant of its natural fize, and alfo when magnified with powers 2 and 1 of his microfcope.

66. *rubiginofa.* C. filis ramofiffimis, rigidis, erectiufculis, rubiginofis; in maffam fub-folidam implexis; articulis diametro fub-quadruplo longioribus. T. 68.

67. *phofphorea.* C. filis ramofis, adfcendentibus, breviffimis, in cruftam uniformem denfiffimè implexis, violaceis; articulis diametro fub-fefqui longioribus. T. 88.

68. *purpurea.* C. filis dichotomis cœfpitofis, implexis, minutiffimis, faftigiatis, purpureis; dichotomis approximatis; articulis diametro fub-duplo longioribus. T. 43.

69.* *lichenicola.* C. filis ramofis cœfpitofis, abbreviatis, aureis, ficcitate demum cinereis; ramis longis alternis; articulis torofis, diametro fub-duplo longioribus.

*C. lichenicola. Eng. Bot.* t. 1609.

On Lichens. In the New Foreſt; *Mr. Lyell.* About Belfaſt; *Mr. Templeton.* In Houghton and St. Leonard's Foreſt, Suſſex; *Mr. Borrer.*

This ſpecies is nearly related to C. *aurea*, from which ſome of the filaments ſeem ſcarcely to differ, except in their ſmaller ſize.

70. *aurea.* C. filis ramoſis cœſpitoſis, abbreviatis, aureis ſiccitate demum cinereis; ramis longis patentibus rigidiuſculis, ſub-incurvis; articulis cylindraceis, diametro ſeſquilongioribus. T. 35 and T. C.

Since the publication of my deſcription of C. *aurea* I have diſcovered it with capſules, which are repreſented in the ſupplementary Plate C. magnified 1. *C. ilicicola* of Engliſh Botany does not appear to me at all diſtinct from this ſpecies, and I have been favored by Mr. Templeton with ſome ſpecimens gathered on the trunks of Quercus Ilex, in Lord Dungannon's Park, near Belvoir, in Ireland, with capſules preciſely reſembling thoſe of *C. aurea.*

71.* *olivacea.* C. filis ramoſis, erectis, cœſpitoſis, implexis, abbreviatis, rigidiuſculis, fuſco olivaceis; ramis ſubſimplicibus, alternis, obtuſis; articulis longitudine diametrum æquantibus. T. C.

On Marine Rocks in Papa Weſtra, Orknies. *Mr. Borrer* and *Mr. Hooker.*

I am indebted to Mr. Borrer for ſpecimens of this hitherto nondeſcript ſpecies, which, in company with Mr. Hooker, he diſcovered during their late tour through Scotland. The filaments of a browniſh olive color, are not more than a quarter of an inch in length, and grow ſo matted together as to form a minute turf on the rocks. It may be diſtinguiſhed from *C. radicans*, to which it ſeems moſt nearly allied, by its different mode of growth, ſhorter filaments and longer joints. The drawing was made by Mr. Hooker from a dried ſpecimen, and repreſents a filament when magnified with powers 3, 2 and 1 of his microſcope.

72.* *radicans.* C. filis ramoſis hic illic radicantibus, ſtrictis, rigidiuſculis, fuſco olivaceis; ramis ſimplicibus, ſparſis, erectis, obtuſis, baſi attenuatis; articulis diametro ſub-dimidio brevioribus. T. C.

On ſandy Banks among the Rocks in Bantry Bay; *Miſs Hutchins.* Rocks at Hartlepool; *Mr. Backhouſe.*

Miſs Hutchins firſt diſcovered this ſpecies of Conferva in the neighbourhood of Bantry, and the preſent deſcription is made from a drawing and ſpecimens which ſhe ſent to Mr. Turner. The filaments grow to the length of about half an inch, and according to Miſs Hutchin's obſervations throw out fibrous roots towards their baſe. The color is of a browniſh olive: the branches, which are erect and diſpoſed without order, are uniformly ſimple with obtuſe apices. The joints are about equal to half of the diameter. The fructification is in capſules which are moſtly ſeſſile, numerous, and diſpoſed on the filaments without order. The ſubſtance is rather ſtiff and not in the leaſt gelatinous, ſo that in drying it adheres to neither glaſs nor paper. The drawing at plate C was made by Miſs Hutchins from the recent plant, and repreſents it both of its natural ſize and when magnified 3, to which Mr. Hooker, from a dried ſpecimen, has added a piece of a filament magnified 1.

73. *Brownii.* *C.* filis ramoſis, densè cœſpitoſis, rigidiuſculis, abbreviatis, viridibus; ramis ramuliſque ſub-ſecundis; articulis apice plerumque incraſſatis, diametro ſub-quintuplo longioribus. T. *D.*

On Wet Rocks in a Cave near Dunrea, Ireland. *Mr. Robert Brown.*

This plant I introduce entirely on the authority of Mr. Brown, who conſiders it as a diſtinct ſpecies, and to whoſe judgment in all matters relating to Botany, the greateſt deference is due. He alone has obſerved it, and I have a pleaſure in publiſhing it with his name. The following deſcription was made by Mr. Brown from recent ſpecimens. "In cœſpitibus denſis nunc convexis nunc planiuſculis latioribuſque. Filamenta (quaſi faſciculata) erecta, ramoſiſſima, 1½ ad 2 lineas longa, craſſiuſcula, rigidula; ramis ſub-ſecundis, dichotomi; articulis multoties longioribus quam latis, pluribus apicem verſus ſenſim incraſſatis, paucis cylindricis. Fructificatio nulla viſa." The ramifications and joints are ſo nearly ſimilar to thoſe of C.

*ægagropila* that I apprehend it can only be diſtinguiſhed from that ſpecies by its very different mode of growth. For the drawing which repreſents *C. Brownii* of the natural ſize, and when magnified 3 and 1, I am indebted to the kindneſs of my friend Joſeph Woods.

74.* *cryptarum.* C. filis dichotomo-ramoſis, repentibus, viridibus; ramis divaricatis acuminatis, articulis diametro ſub-triplo longioribus. T. *D.*

In Caves. North of Ireland; *Mr. R. Brown.* In the firſt Cave on the Cave Hill near Belfaſt, growing among *Hypnum tenellum; Mr. Templeton.* In Caves by the Sea-ſide near Bantry; *Miſs Hutchins.*

Mr. R. Brown, who firſt diſcovered this plant ſeveral years ago, favored me with a ſpecimen under the preſent name. It is of about the ſize of *C. velutina*, but its mode of ramification is widely different. The magnified drawing at plate *C* was made from a dried ſpecimen by *Mr. Hooker.*

75. *velutina.* C. filis ramoſis, repentibus, abbreviatis, pulvinatis, implexis, lœtè viridibus; ramis erectis obtuſis; articulis diametro multuplo longioribus. T. 77.

76. *umbroſa.* C. filis ramoſis repentibus, abbreviatis, fragilibus, nigro viridibus; ramis curvis, ſimplicibus, ſub ſecundis, obtuſis, articulis cylindraceis inflatiſque longitudine variantibus. T. 61.

77. *multicapſularis.* C. filis ramoſis, repentibus, nigro-olivaceis; ramis erectis, ſimpliciuſculis, brevibus, apicem verſus incraſſatis & capſuliferis; capſulis congeſtis, articulis longitudine variantibus. T. 71 and T. *D.*

At Plate *D* is repreſented an extraordinary appearance of this ſpecies, which I have obſerved ſince my deſcription was publiſhed in a ſpecimen gathered near Swanſea. The drawing was made by *Mr. Young.*

78. *pulveria.* C. filis dichotomo-ramoſis, repentibus, minutiſſimis, apice capſuliferis, æruginoſis; diſſepimentis ſub-obſoletis, articulis diametro triplo longioribus. T. *D.*

*Byſſus æruginoſa. Fl. Ang.* p. 605. *Withering.* IV. p. 143.

On the Stems of dead Fern; *Col. in Dillenius.* On rotten Wood;

*Hudſon.* On the Pillars of Roſlyn Chapel near Edinburgh; *Dr. Smith.* On the Ruins of the Chapter Houſe at Margam, and the Walls of Oyſtermouth Caſtle, Glamorgan; *Mr. Young.*

This ſpecies, for the diſcovery of which I am indebted to Mr. Young, ſo nearly accords with the deſcription in the Hiſtoria Muſcorum, that I feel no heſita ion in publiſhing it as the *Byſſus æruginoſa* of Hudſon. It is an extremely minute ſpecies, of a bluiſh green color, and rather powdery appearance. When examined with the higheſt powers of the microſcope, the filaments are ſeen to be twice or thrice dichotomous, and diſſepiments may be here and there obſerved, dividing them into joints, whoſe length is about equal to three times their diameter. Mr. Young remarked that the branches are ſometimes ſingularly reflected. On the termination of each branch there are generally two oval bodies of a dark green color, which I ſuppoſe are either capſules or naked ſeeds, but they are ſo minute that it is impoſſible to ſpeak with any certainty of their nature, and it is theſe which give the plant its powdery appearance.

The drawing at Plate D was made by Mr. Young, and repreſents the plant both of its natural ſize and when magnified with the higheſt power of a compound microſcope.

79. *ebenea.* C. filis ramoſis, erectis, abbreviatis, cartilagineis, nigris; ramis ramuliſque obtuſis; articulis diametrum longitudine æquantibus. T. 101.

80. *atro-virens.* C. filis ramoſis, rigidiuſculis, atro-virentibus; ramis ſub-ſecundis utrinque attenuatis; articulis breviſſimis tripunctatis. T. 25 and T. D.

*Lichen exilis.* AUCTORUM.

The fructification of this ſpecies, which I diſcovered on ſome ſpecimens gathered near Beddgellart, is repreſented at plate D magnified 1.

81.* *ocellata.* C. filis ramoſis, flaccidis, intra moniliformibus, fuſceſcentibus; ramis ſub-ſecundis, remotis elongatis, ſimplicibus; articulis diametro dimidio brevioribus, centro ſæpè notatis. T. D.

On a Bog on Town Hill Common, near Southampton. *Joseph Woods, Junr. Esq.*

I am obliged to Mr. Woods for the ſketch and ſpecimens from which I have taken this deſcription, and which are the only ones I have ever ſeen of this ſingular ſpecies. The filaments do not appear to poſſeſs any real diſſepiments, but a chain of bead-like globoſe veſicles, conſiderably narrower than themſelves paſs through them, in the center of moſt of which another concentric veſicle may be obſerved. The filaments ſometimes, like thoſe of *C. atro virens*, are not of the ſame thickneſs thoughout, and with this ſpecies *C. ocellata*, though extremely different, ſeems to poſſeſs moſt affinity. The color to the naked eye is brown, but under the microſcope, when examined with a ſtrong light, appears almoſt of an orange hue. The figure at plate D repreſents the plant as it appears when magnified with powers 2 and 1.

82. *caſtanea.* C. filis ramoſis, repentibus, pinnatis, acuminatis, caſtaneis; pinnis pinnuliſque alternis, divaricatis; articulis caulis longiſſimis, pinnarum brevioribus. T. 72.

Mr. Turner is of opinion that this is the *C. muſcicola* of the German authors, but it does not well accord with the figure in Weber and Mohr's *Reiſe durch Schweden*, or Dr. Roth's deſcription in the Catalecta Botanica.

83. *Acharii.* C. filis ramoſis, cœſpitoſis, rigidiuſculis, ſub-erectis, fuſco-olivaceis; ramis brevibus, patentibus, apicibus obtuſis; articulis diametro ſub-duplo longioribus. T. 89.

84. *orthotrici.* C. filis ramoſis, cœſpitoſis, pulvinatis, rigidiuſculis, fragilibus, obtuſis, caſtaneis; ramis ſub-alternis; articulis diametro vix longioribus. T. 89.

85. *chalybea.* C. filis ramoſis, pulvinatis, faſtigiatis, ſtrictis, tenuibus, erectis, nigro-viridibus; ramis ſub-alternatim ſecundis; ramulis lateralibus, breviſſimis, multifidis, capſuliferiſque; articulis diametro quintuplo longioribus. T. 91.

*C. corymbifera.* *Eng. Bot.* t. 1996.

Since this plant was defcribed in my work, Mr. Backhoufe has found fome fpecimens of it near Darlington, as large as thofe fent by Dr. Roth to Mr. Turner, and exactly agreeing with them, (as well as with thofe from Mr. Borrer, excepting only in the greater length of their filaments). Upon thefe the fructification was firft difcovered, which is fo remarkable and fingular that Dr. Smith was mifled by it to regard Mr. Backhoufe's plant as a new fpecies, and to publifh it as above quoted in Englifh Botany.

86. *vivipara.* C. filis dichotomo ramofis, flexuofis flavo virentibus; ramis ad diffepimenta bubbiferi; hulhis piliferis; articulis diametro triplo longioribus. T. 59.

*C. fetigera.* ROTH. *Cat. Bot.* III. p. 283. t. 8. f. 1.

Since I publifhed the defcription of this fpecies it has been found in the neighbourhood of Darlington by Mr. Backhoufe.

87.* *exigua.* C. filis ramofiffimis, minutis, gelatinofis, viridibus; ramis confertis; ramulis elongatis apice pellucidis; articulis diametrum longitudine fub-æquantibus. T. *D.*

In the Chalybeate Stream which runs through the Bog on Apfe Heath, near Shanklin, Ifle of Wight. *J. Woods, junr. Efq.*

I received a fpecimen and drawing of this minute and beautiful fpecies from my friend Mr. Woods, who informs me that its length is not greater than three fixteenths of an inch. The length of the joints in the principal branches fomewhat exceeds the diameter, but thofe of the ramuli are fhorter. This fpecies feems nearly related to the *Rivulariæ.*

88. *protenfa.* *C.* filis ramofis, lubricis, viridibus; ramis diffufis, maximè elongatis, apice pellucidis; articulis diametro fub-fefquilongioribus. T. 67.

89. *lubrica.* C. filis ramofiffimis, lubricis, viridibus; ramulis fparfis, approximatis, aculeiformibus; articulis diametro faltem triplo longioribus. T. 57.

90. *mutabilis*. C. filis ramofiffimis, fubmoniliformibus, gelatinofis, viridibus; ramulis fafciculatis, multifidis, penicilliformibus, apice protenfis; articulis diametro fefquilongioribus. T. 12.

91. *gelatinofa*. C. filis ramofiffimis, moniliformibus, gelatinofis, obfcurè viridibus; ramulis fubverticillatis, multifidis, penicilliformibus; articulis ramulorum longitudine diametrum fub æquantibus. T. 32.

Since I publifhed my defcription of this fpecies, I have difcovered the blue variety in Llyn Cwellyn, and examined it carefully on the fpot with a compound microfcope. The principal ftems were entirely deftitute of whirls, but the ends precifely refembled thofe of the plant in its common ftate, the color alone excepted. I am inclined to think that the fingular appearance of this variety arifes from difeafe, probably occafioned by its alpine fituation, and its growth in ftagnant water, but at all events it has no claim whatever to be confidered a diftinct fpecies.

92. *atra*. C. filis ramofiffimis, moniliformibus, fub-gelatinofis, atro-viridibus; ramulis fetaceis; articulis diametro quintuplo longioribus, fupernè incraffatis, verticillato-ciliatis. T. 11 and T. *D*.

Since the publication of this fpecies it has alfo been found at Bantry by Mifs Hutchins, and near Cambridge by Mr. Relhan, who firft difcovered the fructification, which has been fubfequently found about Yarmouth by Mr. Turner and Mr. Hooker. The fruit is large, globular, and feffile, of a dark color, and fcattered plentifully over the frond, efpecially near the bafe, in which refpect it differs from moft other Confervæ. Although the powers of my microfcope did not enable me fatisfactorily to determine, yet I have but little doubt that thefe capfules refemble in their nature thofe of C. *gelatinofa*; they are reprefented in my fupplementary plate *D* magnified 1.

93.* *nigricans*. C. filis dichotomis, rigidiufculis, viridi-nigricantibus; ramis longis, remotis, patentibus; articulis diametro quadruplo longioribus. T. *E*.

C. *nigricans*. Roth. *Cat. Bot.* III. p. 277.

In a pond at Wimbledon, Surry. *Mr. Dickfon.*

Mr. Dickfon alone appears to have difcovered this fpecies in Britain, and to him I am indebted for the fpecimens now before me, which having been fent by Mr. Turner to Dr. Roth were returned with the name of *C. nigricans.* The filaments grow to the length of three or four inches, and are irregularly divided by patent dichotomies. The joints, whofe length is about four times greater than the diameter, are by Dr. Roth defcribed "fporulis ubique fparfis," and in the fpecimen now before me moft of them are covered by dark colored fpots, which however feem rather to proceed from decay or fome extraneous matter attached to them. In drying it will not in the leaft adhere to either Glafs or Paper. In Plate E the plant is reprefented when magnified 4 and 2, and I am indebted to my friend Mr. Hooker for the drawing, which was neceffarily made from a fpecimen that had been previoufly dried.

94. *crifpata.* C. filis ramofis, crifpatis, faturatè viridibus; ramis alternis, remotiffimis; articulis diametro multuplo longioribus, ficcitate alternatim compreffis. T. 93.

95. *pennatula.* C. filis ramofiffimis, flavefcentibus; ramis ramulifque erecto-patentibus, fub-incurvis; articulis cylindraceis diametro fextuplo longioribus.

C. *pennatula.* Fl. Dan. t. 945.

Ditches about Yarmouth.

Of this plant I have now no fpecimens, but a drawing made by myfelf in 1802, from fome individuals gathered near Yarmouth, fo exactly accords with the figure in the *Flora Danica,* that I am led to admit it as a Britifh Conferva, though chiefly for the purpofe of directing the attention of other Botanifts to the fubject, and without by any means pledging myfelf for its being a diftinct fpecies.

96.* *flavefcens.* C. filis ramofiffimis, flexuofis flavo virentibus; ramis fub-dichotomis; patento-horizontalibus; ramulis lateralibus abbreviatis; articulis cylindraceis diametro decuplo longioribus. T. *E.*

*C. flavescens.* Roth. *Cat. Bot.* II. p. 224. III. p. 241. *Fl. Germ.* III. parf. 1. p. 511.

In the Ditches at Cley, Norfolk; *Mr. Hooker.* In the New River at Stoke Newington; *Mr. Woods.*

This fpecies, though nearly allied to *C. fracta*, is diftinguifhable by its more flender filaments and by its longer joints. The drawing at plate D was made by Mr. Hooker, and reprefents the plant magnified 3 and 2.

97. *fracta.* C. filis ramofiffimis flexuofis viridibus; ramis ramulifque divaricatis fub-alternis; articulis diametro quintuplo longioribus demum oblongis. T. 14.

This is a very variable fpecies, fo much fo that in particular fituations it approaches fo clofely both to the preceding and following one as to require fometimes great care to diftinguifh them. It has been already noticed in the Introduction that it is the C. *vagabunda* of Hudfon.

98. *flexuofa.* C. filis dichotomo-ramofis, rigidiufculis, faturatè viridibus; ramis flexuofis; ramulis fub-fimplicibus, tenuiffimis, alternatim fecundis, patentibus, articulis diametro duplo longioribus. T. 10.

99.* *Hutchinfia.* C. filis ramofiffimis, flexuofis, fub-cartilageneis, fragilibus, glauco viridibus; ramis ramulifque fparfis, ultimis fecundis adpreffis; articulis torolofis, diametro duplo longioribus. T. 109.

100. *diffufa.* C. filis dichotoma-ramofis, flexuofis, rigidis, viridibus; ramis diffufis remotis; ramulis brevibus approximatis, obtufis; articulis cylindraceis, diametro quadruplo longioribus. T. 21.

101. *rupeftris.* C. filis ramofiffimis, ftrictis, virgatis, fafciculatis, intensè viridibus; ramis adpreffis, obtufis; articulis cylindraceis, diametro fub-quadruplo longioribus. T. 23.

102. *glomerata.* C. filis ramofiffimis, rigidiufculis, viridibus; ramis alternis; ramulis brevibus, fecundis, fub faftigiatis, penicilliformibus, obtufiufculis; articulis diametro quadruplo longioribus. T. 13.

103. *læte-virens.* C. filis ramofiffimis, rigidiufculis, arcuatis, læte viridibus; ramis approximatis, acuminatis; ramulis brevibus, alternatim fecundis; articulis diametro fub-triplo longioribus. T. 48.

104.* *albida.* C. filis ramofiffimis, coacervatis, tenuibus, albo-virefcentibus; ramis fubquaternis approximatis; ramulis horizontalibus, oppofitis, flexuofis, ultimis fub-fecundis; articulis diametro quadruplo longioribus. T. *E.*

C. *albida.* *Fl. Ang.* p. 595. WITHERING. IV. p. 131.

C. *marina tomentofa, tenerior, & albicans.* DILL. *Hift. Mufc.* p. 19. t. 3. f. 12.

β. *protenfa.* Fills in longum protenfis; ramulis patentibus, ftrictis.

In the Sea at Cromer; *Mr. Turner.* β. Coaft of Suffex; *Mr. Borrer.* In Bantry Bay, not uncommon during the months of June and July; *Mifs Hutchins.*

By means of a fragment of the Dillenian fpecimen No. 12, I have been enabled to fatisfy myfelf that the prefent is the fame fpecies, and confequently the C. *albida* of Hudfon. I have at prefent feen only two dried fpecimens of this plant, the habit of which is fo remarkably thick and cluftered that it is extremely difficult to extricate a fmall piece fo as clearly to difcover the ramification. Its ftrongeft character feems to lie in the oppofite and horizontal ramuli. The color is a pale waxy yellowifh green, wholly devoid of glofs. The length of the filaments is about three inches. The variety β is feven or eight inches long, and of a lefs bufhy habit. Its ramuli are lefs regularly oppofite, and are ftrait inftead of being flexuofe. Mifs Hutchins fays that when frefh it is of a beautiful pale green color. For the fketch at Plate E, which reprefents the ramuli magnified 2, I am indebted to Mr. Hooker.

105. *pellucida.* C. filis ramofiffimis, ftrictis, rigidis, dilute viridibus; ramis plerumque ternis, obtufis; articulis longiffimis. T. 90.

106. *ægagropila.* C. filis ramofiffimis, viridibus, e centro progredientibus, et globum conftituentibus; ramis ramulifque fub-fecundis, obtufis; articulis diametro quadruplo longioribus. T. 87.

107.* *æruginosa.* C. filis ramosis, flexuosis, brevibus, æruginosis; ramis sparsis, patentibus, obtusis; articulis diametro sub-sesquilongioribus. T. *E.*

C. *æruginosa.* *Fl. Ang.* p. 595. WITH. IV. p. 131.

*C. marina capillacea brevis, viridissima mollis.* DILL. *Hist. Musc.* t. 4. f. 20.

On Fuci.

The sketch of this Conferva represented in the supplementary Plate E, as also the above description, is taken from the original specimen in the Dillenian Herbarium, and is published because I have seen no other British specimen that resembles it. I have neither gathered it myself, nor ever seen it in any other collection. It is from half to three-quarters of an inch in length. The drawing represents a filament magnified 1.

108.* *arcta.* C. filis ramosis, strictis, virgatis, cœruleo-viridibus; ramis sub-patentibus, ultimis sparsis adpressis; articulis inferioribus, brevibus, superioribus, longissimis. T. *E.*

In the Sea, Bantry Bay. *Miss Hutchins.*

My friend Mr. Turner favored me with specimens of this species, which he received from Miss Hutchins, to whom the botanical world is indebted for its discovery. It grows to the length of two or three inches, and is of a light bluish green color. The filaments are about twice divided: the branches issue at acute angles and at uncertain distances from each other; they are most commonly alternate but sometimes opposite, and a few of those near the root, in the specimen now before me, contrary to their general character, are curled inwards in a remarkable manner. The length of the joints varies; in the lower part of the filament it scarcely exceeds the diameter, but becomes longer towards the summit, and the terminal joints are remarkably long. When dried, in which state alone I have hitherto had an opportunity of observing it, it has a flaccid Ulva like appearance. For the drawing at Plate E, I am indebted to Mr. Hooker: the plant is represented magnified 4, and also the lower and upper end of a filament, separately, magnified 2.

109.* *lanosa*. C. filis ramosis, brevibus, tenuibus, luteo-virescentibus; ramis sparsis; articulis inferioribus sub-duplo, ultimis multuplo diametro longioribus. T. *E*.

*C. lanosa*. ROTH. *Cat. Bot.* III. p. 291.

β. *Zosteræ*. Filis læte viridibus, splendentibus.

On Rocks and Algæ in the Sea. Near Forres; *Mr. Brodie*. At Cromer; *Mr. Hooker*. Anglesea; *Rev. H. Davies*. At Brighton; *Mr. Borrer*. At Ilfracombe; *Miss Hill*. Bantry Bay; *Miss Hutchins*. Between Dover and the South Foreland. β. On Zosteræ at Worthing; *Mr. Borrer*. On Marine Algæ, near Forres; *Mr. Brodie*.

That this species is the C. *lanosa* of Roth, I have been enabled to prove by means of authentic specimens in Mr. Turner's extensive Herbarium. I discovered it several years ago in the neighbourhood of Dover, and have since received specimens from several of my friends. The filaments are mostly about four lines, and I believe they never exceed an inch in length. The color is generally of a very dull yellowish green, wholly destitute of gloss when dry. The joints vary in length, some of those in the lower part of the filament being about equal to, and others double the diameter, but those at the terminations of the filaments are uniformly much longer than any of those below them. Mr. Hooker in the specimens which he gathered at Cromer, observed two small dark colored spots in many of the joints, but this appearance they lose in drying. The var. β. was sent me by Mr. Botrer, who found it on the Sussex coast growing on Zostera marina, and I have also received it from Mr. Brodie: it is of a grass-green color and is glossy, but though on this account widely different at first sight, it does not appear to be distinct from the present species. Mr. Turner has received both of these from Miss Hutchins as the same. The drawing at plate E was made by Mr. Hooker from a recent specimen, and represents the plant magnified 3, and also separately the upper and lower parts of a filament magnified 1.

110. *tomentosa*. C. filis ramosissimis, tenuibus, funis in formam densissimè contortis, sub-ferrugineis; ramis divaricatis, ultimis simplicibus; articulis diametro quadruplo longioribus. T. 56.

111.* *riparia*. C. hiis infernè simpliciusculis, supernè ramosis, longis tenuibus, implexis flavo-virentibus; ramis remotis, divaricatis; articulis diametro sub duplo longioribus. T. *E*.

C. *riparia*. ROTH. *Cat. Bot.* III. p. 216.

Near Bantry; *Miss Hutchins*. In Salt pools by the Yare, near Yarmouth.

It is on the authority of authentic specimens in Mr. Turner's Herbarium, which I have compared with those sent by Miss Hutchins, that I publish this species with the reference to Roth, upon whose description I have been under the necessity of relying for a part of my own, the filaments being so long and entangled that in a dried specimen it is almost impossible to separate them. I have a sketch which belongs to the same species, and which I made from a plant that I discovered in pools by the side of the Yare, near Yarmouth, in 1802. The filaments towards the root have but few branches, but they are more numerous towards the summits, and always remarkably divaricated. The drawing at T. *E*. was made by Mr. Hooker from a dried specimen, and represents, separately, the ramification and nature of the joints, magnified with powers 2 and 1 of his microscope.

112.* *siliculosa*. C. filis ramosissimis, tenuibus, fusco-flavescentibus; ramis ramulisque sub alternis, acuminatis; articulis diametrum longitudine æquantibus; capsulis siliculiformibus. T. *E*.

*Ceramium confervoides*. ROTH. *Cat. Bot.* I. p. 151. t. 8. f. 3. III. p. 148. *Fl. Germ.* III. pars I. p. 467.

Rocks in the Sea at Cromer and Hastings. *W. J. Hooker, Esq.*

Dr. Roth and my friend Mr. Hooker are of opinion that *C. siliculosa* is specifically distinct from C. *littoralis*, to which I have thought it right to accede, never having myself had the opportunity of comparing recent ſpecimens of the two plants together. The principal difference which I can diſcover, conſiſts in the lanceolate pods of the one, contraſted with the globular capſules of the other, but this however I can hardly admit to be a ſufficient indication of ſpecific difference, ſince the ſame may be obſerved between C. *coccinea* and its variety, in C. *arbuſcula*, and ſeveral other Confervæ, each of which ſhould in that caſe be divided into two ſpecies. The drawing at Plate E was made by Mr. Hooker from the recent plant, and repreſents the filaments magnified 3 and 1.

113. *littoralis*. C. filis ramoſiſſimis, tenuibus, implexis, olivaceis; ramis ramuliſque ſub-alternis, acuminatis; articulis diametrum longitudine æquantibus; capſulis globoſis. T. 31.

It appears from the third faſciculus of the Catalecta Botanica, that Dr. Roth's *Ceramium tomentoſum* belongs to the preſent ſpecies, and is quite different from C. *tomentoſa* of Hudſon, to which I had erroneouſly referred it. The latter is probably *Ceramium compactum* of Roth.

114. *fœtida*. C. filis ramoſis, coadunatis, virgatis, apicibus liberis, olivaceis; ramis confertis; articulis diametro ſeſqui longioribus, granula elliptica includentibus. T.

115.* *paradoxa*. C. filis ramoſis coadunatis, tenuiſſimis, lubricis, dilutè viridibus; ramis longis ſparſis, adpreſſis; articulis diametrum longitudine æquantibus, granula ſphærica includentibus. T. *F.*

In the Sea at Bangor; *Mr. Templeton.* Beach at Brighton; *Mr. Borrer.*

A ſpecimen from Mr. Templeton in Mr. Turner's Herbarium, proves that he was the firſt diſcoverer of this moſt extraordinary ſpecies. It has alſo been gathered on the Suſſex coaſt by Mr. Borrer, and it is through his aſſiſtance that I am enabled to offer the following obſervations reſpecting it. It grows in cloſe tufts four or five inches long; the color of my dried ſpecimens is light green, but in the place of growth it is probably different, as Mr. Botrer in thoſe which he picked up on the beach at Brighton obſerved a purple tinge, and was thereby led to ſuſpect that they had ſuffered ſome change in this reſpect. It is irregularly and repeatedly divided with branches, which are moſtly oppoſite, but often alternate and not unfrequently crowded together. The ultimate ramuli are very long. What to the naked eye appears to be a ſingle filament, under the higher powers of the microſcope, is ſeen to conſiſt of many agglutinated, or adhering cloſely together in the ſame manner as thoſe of *C. fœtida*, with which I apprehend this ſpecies poſſeſſes conſiderable affinity. Each individual of theſe extremely ſlender filaments is ſeparately jointed. The length of the joints is about equal to their diameter, and ſo far as I am able to judge from a dried ſpecimen, they each include a globule, of the ſame nature with thoſe of C. *fœtida*. The ſketch at Plate F (made by Mr. Hooker from a dried ſpecimen magnified 4 and 2) will ſerve to convey ſome idea of the plant, but I apprehend that it ſuffers more than moſt other ſpecies in drying, and it is principally from the obſervations of Mr. Borrer that this deſcription has been made.

116. *nana.* C. filis ramoſis, minutiſſimis, fuſco viridibus; ramis ramuliſque ſub alternis acuminatis; articulis diametro duplo longioribus. T. 30.

117. *minutiſſima.* C. filis ſub-ramoſis, minutiſſimis, hyalinis; ramis ſparſis, furcatis, obtuſiuſculis; diſſepimentis obſoletis; articulis longitudine variabilibus. T. F

On Confervæ in the Sea.

This ſpecies, which has been obſerved both by Mr. Borrer and myſelf growing paraſitically on ſeveral of the Marine Confervæ, is ſo extremely minute as to be nearly imperceptible to the naked eye, and even the higheſt power of my microſcope is hardly ſufficient to aſcertain its nature. The filaments are ſometimes ſimple, but have moſt uſually two or three branches which are frequently forked. Diſſepiments may now and then be faintly diſtinguiſhed at uncertain diſtances from each other, but with this exception no mark of internal organization or even color can be obſerved. For the drawing at Plate F, which repreſents the plant magnified 2 and 1, I am indebted to Mr. Woods.

118. *lanuginosa*. C. filis sub-ramosis, minutissimis, ferrugineis; ramis sparsis, obtiusiusculis; articulis medio pellucidis, diametro triplo longioribus. T. 45.

119.* *pluma*. C. filis repentibus, ramosis, minutis, intense roseis; ramis erectis infra denudatis, superne pinnatis pinnis oppositis, approximatis; articulis diametro duplo longioribus. T. F.

On the stalks of Fucus digitatus in Bantry Bay. *Miss Hutchins.*

This beautiful species, of which a drawing and specimens were communicated by Miss Hutchins to Mr. Turner, may be readily distinguished from C. *repens* and *C. tenella*, to which it is most nearly allied, by having the erect branches thickly pinnated with opposite ramuli towards their apices. The capsules are globose and mostly terminal. The drawing at Plate F, for which I am indebted to Mr. Hooker, represents *C. pluma* of the natural size, and also when magnified 1.

120. *repens*. C. filis repentibus, ramosis, implexis, minutis rufis; ramis erectis; ramulis sub secundis obtusis; articulis diametro triplo longioribus. T. 18.

My former reference to Dillenius is erroneous, as has been pointed out by Mr. Turner in his remarks on the Dillenian Herbarium, *Lin. Transf.* VII. p. 106.

121. *tenella*. C. filis repentibus, ramosis, implexis, minutis dilute roseis; ramis creetls, simplicibus; articulis longitudine variabilibus. T. *F*.

On the Shells of the large Scallop at Bantry. *Miss Hutchins.*

The present is one of the numerous species for the discovery of which the botanical world is indebted to *Miss Hutchins*. The filaments are of the same size and strike root precisely in the same manner as those of C. *repens*, from which it differs in its lighter color, extremely flaccid nature, more slender growth, and in having the erect branches undivided. The drawing at Plate F was made by Mr. Hooker from a dried specimen, and represents *C. tenella* both of the natural size and when magnified 1.

122.* *Daviesii.* C. filis ramosis, erectis minutis, liberis roseis; ramis sparsis acuminatis; articulis diametro triplo longioribus. T. F.

On Marine Algæ; *Rev. Hugh Davies.* Bantry Bay; *Miss Hutchins.* At Brighton; *Mr Borrer.*

I have a pleasure in naming this species after my valuable friend, the Rev. Hugh Davies, whose intimate knowledge of many branches of Natural History is well known, and to whose liberality this work is greatly indebted. Its length rarely exceeds three or four lines, and it may be distinguished from its congeners by its unentangled growth, and far different ramification. Mr. Borrer informs me that he has once discovered it with capsules, placed in rows along the upper side of the ramuli. For the drawing at Plate E. I am indebted to Mr. Hooker; it represents the plant magnified 3, and a piece of the filament magnified 1.

123. *Rothii.* C. filis dichotomo-ramosis, erectis, brevibus, densè cæspitosis, phœniciis; ramis ramulisque alternis; articulis diametro sub-triplo longioribus. T. 73.

Several years ago I received a specimen of this plant from Mr. Robert Brown, gathered by himself in the North of Ireland, and which he had named *C. phœnicia.* It was not till after I had published my description of *C. Rothii* that I recognised it as the same species, which I much regret, as Mr. Brown certainly first discovered it in Britain. It has since been found by Mr. W. W. Young, near Dunraven, in Glamorganshire, and by Messrs. Hooker and Borrer, on the Coast of Durness, Sutherland.

124.* *floridula.* C. filis ramosis, tenuibus, cæspitosis, implexis, dilutissimè roseis; ramis sparsis, simpliciusculis, remotis; articulis diametro sub-triplo longioribus. T. F.

Rocks on the Sea shore. On the Galway Coast; *Dr. Scott.* On the Antrim Coast; *Mr. Mackay.*

I received specimens of this species from the late Dr. Scott, gathered on the Galway coast, where it covers the rocks on the Sea shore. The filaments are much finer than human hair, but their growth is ſo entangled, that in a dried ſpecimen it is almoſt impoſſible to ſeparate them ſo as to aſcertain their length, which is I believe generally about half an inch; when freſh, according to Mr. Mackay's obſervations, they are of a fine bloom color, but this they loſe in drying and then become of a reddiſh dull green. The ſketch at Plate F was made by Mr. Hooker from a dried ſpecimen, and repreſents the filament magnified 3 and 1.

125.* *interrupta.* C. filis ramoſis, breviuſculis, purpuraſcentibus; ramis ramuliſque alternis; articulis ſurſum incraſſatis, truncatis, diametro ſub-quadruplo longioribus.

*C. interrupta. Eng. Bot.* t. 1838.

On Marine Confervæ. At Brighton; *Mr. Borrer.* In Bantry Bay; *Miſs Hutchins.*

The capſules of this ſpecies are divided in a remarkable manner by a tranſverſe pellucid line.

126. *pedicellata.* C. filis dichotomo-ramoſis rubris; ramulis alternis multifidis; articulis ſurſum incraſſatis, diametro ſub-quintuplo longioribus. T. 108.

127. *ſetacea.* C. filis dichotomo-ramoſis, virgatis, ſtrictis, intensè ſplendidèque roſeis; ramis elongatis; articulis ſub cylindraceis diametro ſub quintuplo longioribus. T. 82.

128. *corallina.* C. filis dichotomis, lubricis, ſplendidè aureo-rubris; articulis ſurſum incraſſatis, diametro quadruplo longioribus. T. 98.

Since I publiſhed my deſcription of C. *corallina* I have not ſeen any recent ſpecimens, but I have examined many in a dried ſtate, and theſe have led me more and more to ſuſpect that my drawing, as well as former obſervations reſpecting the fructification of this ſpecies, are in at leaſt ſome degree inaccurate. It is impoſſible to form any decided opinion from dried ſpecimens, but I am inclined to believe that the involucrum, till the ſeeds have arrived at maturity, ſo cloſely and compactly envelop the internal jelly, as to bear the refemblance as well as anfwer the purpofe of a capfule. In fome fpecimens I have feen a ftill ftronger refemblance of capfules, than what I have figured at D, but they were of a fmaller fize, and had evidently not arrived at maturity, which having attained, they would I apprehend by an expanfion of the involucrum, have appeared as is reprefented in Englifh Botany, with their internal jelly expofed without any covering.

129.* *barbata.* C. filis dichotomo-ramofis, læte fanguineis, apice fibrofis; fibris multifidis tenuiffimis; articulis furfum incraffatis, diametro quintuplo longioribus.

C. *barbata.* *Eng. Bot.* t. 1814.

On the Beach at Brighton. *Mr. Borrer.*

The feeds of this fpecies are imbedded in naked jelly, and guarded by an involucrum inftead of a capfule.

130.* *multifida.* C. filis ramofis, rubris; ramulis fub-ternatis, diftantibus, brevibus, multifidis; articulis diametro multuplo longioribus.

C. *multifida.* *Eng. Bot.* t. 1816. (excl. Syn.)

In the Sea. On the Devonfhire Coaft; *Mrs. Griffiths.* On the Beach at Brighton, and near Newhaven; *Mr. Borrer.* In Bantry Bay; *Mifs Hutchins.*

Dr. Smith erroneoufly fuppofed this fpecies to be Hudfon's C. *multifida*, forgetting that as well as C. *imbricata* it had been before proved to be C. *equifetifolia* of Lightfoot. As however *multifida* has never been ufed as a name for C. *equifetifolia*, there cannot I apprehend be any objection to its being retained for the prefent fpecies. Mr. Borror informs me that he has difcovered a fructification on this fpecies, differing from the one reprefented in Englifh Botany, and of the fame nature with that of C. *barbata.*

131. *equifetifolia.* C. filis ramofiffimis, craffis, rubris; ramis utrinque attenuatis, ramulis verticillatis, imbricatis, brevibus, multifidis, undique obfeffis; articulis diametro multuplo longioribus. T. 54.

The Rev. G. R. Leathes difcovered the fructification of this fpecies in a ſpecimen which he gathered in Anguſt, 1807, on the beach at Yarmouth. It is of the ſame nature with that of C. *barbata*, conſiſting of ſeeds immerſed in a pellucid jelly, and ſurrounded by numerous filaments which wholly envelop it. It is ſcattered over the ſides of the branches, and has to the naked eye the appearance of being only very young ſhoots.

132. *verticillata*. C. filis dichotomo-ramoſis, cartilagineis, craſſis, fuſco-olivaceis; ramulis verticillatis, incurvis, breviſſimis, plerumque bifurcis, undique obſeſſis; articulis diametro brevioribus. T. 55.

133. *ſpongioſa*. C. filis ramoſis, cartilagineis, craſſis, olivaceis, ramulis ſimplicibus, breviſſimis, undique imbricatis; articulis diametro ſub-ſeſquilongioribus. T. 42.

134. *villoſa*.† C. filis ramoſis, flaccidis, craſſis, elongatis, flavis; ramis oppoſitis, remotis, ramulis minutis, pinnatis, ſub-verticillatis, undique obſeſſis; articulis diametro dimidio brevioribus. T. 37. and T. *F*.

In September, 1808, the Rev. G. R. Leathes found a ſpecimen on the Yarmouth Beach, on which Mr. Turner has favored me with the following remarks. "The fibres grow as deſcribed in the *Britiſh. Confervæ*, from every 3d, 4th, or 5th diſſepiment, but rather in tufts than in whirls: they are long, ſometimes ſimple, but moſtly three or four times dichotomous, with acute angles; towards their baſes grow on them ſhort oblong dark-brown bodies (whether ſeeds or capſules it is impoſſible from their minuteneſs to determine) cluſtered and ſeſſile, but from the collapſing of the juices, often looking pedunculate. The filaments are ſo obſoletely jointed that it is difficult to ſay, if they are ſo in reality or not, though they look

† I have received ſpecimens of *Fucus aculeatus* and *Fucus ligulatus* from Mr. Backhouſe, which are covered with ſhort ramuli of the ſame nature and appearance with thoſe of *C. villoſa*. In the former I found to my great ſurprize that the aculei are regularly jointed, and that the main filaments, eſpecially towards their extremities, have a ſimilarly jointed internal tube running longitudinally through them, and occupying nearly half of their width. I was particularly ſtruck with the reſemblance of the joints to thoſe of *C. villoſa*, and they fully confirm Mr. Turner's opinion, that there is a ſtrong affinity between theſe two plants.

ſo in drying." In dried ſpecimens theſe bodies hardly appear to belong to the fructification at all. For the drawing at Plate F, which repreſents the ſuppoſed fructification highly magnified, I am indebted to Mr. Leathes.

135. *fluviatilis*. C. filis ramoſiſſimis, cartilagineis, olivaceis; ramis ramuliſque utrinque attenuatis; diſſepimentis verrucoſis; articulis utrinque dilatatis, diametro ſubquintuplo longioribus. T. 29.

For the fructification of this ſpecies ſee Introduction, p. 20.

136. **torulosa*. C. filis ſub-ſimplicibus, nodoſis, cartilagineis, baſi attenuatis, apice ſub incraſſatis, olivaceis; articulis utrinque contractis, diametro ſub triplo longioribus. T. *F*.

C. *toruloſa*. Mohr *in Schrader's Journal for* 1801, p. 324. t. 3. f. 1. 2. Roth. *Cat. Bot.* III. p. 250. *Fl. Germ.* III. pars 1. p. 529.

*C. fluviatis nodoſa, Fucum æmulans*. Dill. *Hiſt. Muſc.* p. 39. t. 7. f. 48.

In Mountain Streams. Near Ludlow; *Dillenius*. Angleſea; *Rev. H. Davies*.

I am ſtill ſomewhat doubtful whether this ſpecies ſhould be conſidered as diſtinct from C. *fluviatilis*, but I have nevertheleſs admitted it here as ſuch, in reſpect to the opinion of the late Dr. Mohr and Dr. Roth, the former of whom in the German tranſlation of this work, ſays, that he has ſeen the two plants growing together, and is convinced they are perfectly diſtinct, to which I have thought it right to accede, never having myſelf enjoyed an equally favorable opportunity for the examination of them. The fructification is ſimilar to that of C. *fluviatilis*. For the drawing at Plate F, I am indebted to Mr. Hooker, in which the joints are repreſented magnified 5, with a tranſverſe ſection of the filament to ſhew the ſeeds magnified 2, and alſo the ſeeds ſeparated and magnified 1.

137. *ciliata*. C. filis dichotomis, apice forcipatis rubris; diſſepimentis verticillatim ciliatis; articulis medio pellucidis, diametro longitudinem vix ſuperantibus. T. 53.

138. *diaphana*. C. filis ramofiffimis apice forcipatis, purpurafcentibus; diffepimentis obfoletis; articulis utrinque torofis, medio pellucidis, diametro fub longioribus. T. 38.

139. *rubra*. C. filis ramofiffimis rubris; ramulis fetaceis, apice furcatis; articulis utrinque attenuatis, centrum verfus pellucidis, diametrum longitudine fub æquantibus. T. 34.

In Mr. Turner's Herbarium there are bleached fpecimens of a light ftraw color, gathered by Mrs. Griffith at Sidmouth, which differ fo much from the common appearance of C. *rubra*, as to have induced both thefe Botanifts to regard them as belonging to a feparate fpecies. They are fcarcely two inches long, and comparatively thin: their fubftance is remarkably thick and cartilaginous, but the leading difference is in the joints, each of which is marked in the center with a dark globular fpot, nearly fimilar to thofe which may be often feen in C. *polymorpha*. The fructification confifts of feeds fcattered through the interior of the ultimate ramuli, but thefe can by no means be confidered as an indication of fpecific difference, fince they have been alfo obferved in many of the other capfuliferous Confervæ. This remarkable appearance of *C. rubra* is reprefented at Plate F, magnified 5 and 3, from a fketch with which Mr. Hooker favored me.

140. *tetragona*. C. filis ramofiffimis rubris; ramulis patento-horizontalibus, bafi attenuatis, apice acuminatis, fafciculatis, brevibus; articulis ovato-cylindraceis, diametro duplo longioribus. T. 65

141. *tetrica*. C. filis decompofito-pinnatis, fufco-rubris, luridis; pinnis pinnulifque alternis, extremis curvatis; articulis diametro fub-triplo longioribus; capfulis folitariis, pedunculatis. T. 81.

Since the publication of this fpecies it has been found abundantly in Bantry Bay, by Mifs Hutchins, and on the Devonfhire coaft by Mr. Griffiths and Mifs Hill.

142. *rosea.* C. filis decomposito-pinnatis, tenuibus, roseis; pinnis pinnulisque alternis; articulis diametro sub-triplo longioribus; capsulis secundis sessilibus. T. 17

Dr. Roth considers the plant which I have figured to be a variety of *Ceramium roseum.* Mr. Turner and Mr. Borrer are of opinion that the plant which grows in the Yare, and which is that figured in *English Botany* is a distinct species, but I apprehend that every difference between them, entirely arises from the growth of the former in the sea, and of the latter in a river, where the water at some states of the tide, of course contains a much less quantity of salt.

143.* *Borreri.* C. filis decomposito-pinnatis, tenuibus, roseis; pinnis pinnulisque alternis, flexuosis, ultimis fastigiatis; articulis diametro sub-duplo longioribus.

*C. Borreri. Eng. Bot.* t. 1741.

Among the rejectament of the Sea at Yarmouth. *Mr. Borrer.*

I have never seen any other than a dried specimen of this plant, and it is therefore perhaps that I am inclined to doubt, whether it ought to be considered as more than a variety of *C. rosea,* which is a very variable species.

144. *Turneri.* C. filis pinnatis, roseis; pinnis oppositis, sub simplicibus; articulis diametro triplo longioribus. T.

145. *plumula* C. filis ramosis roseis; ramis alternis pinnatis; pinnis oppositis, horizontaliter recurvis; pinnulis secundis; articulis diametro sub triplo longioribus. T. 50.

*C. Turneri. Eng. Bot.* t. 1637.

146.* *Mertensii.* C. filis ramosis, flavescentibus; ramis pinnatis; pinnis sub-oppositis brevibus; articulis diametro dimidio brevioribus.

C. *Mertensii. Eng. Bot.* t. 999.

On the Beach at Yarmouth; *Mr. Wigg.* In Bantry Bay; *Miss Hutchins.* Coast of Durham; *Mr. Backhouse.*

147.* *Hookeri.* C. fills ramofiffimis, primariis incraffatis inarticulatifque, pallidè rufo fufcefcentibus; ramulis confertis, abbreviatis, pinnatis, pinnulis alternis articulatis; articulis diametro fefquilongioribus. T. 106

148.* *arbufcula.* C. filis primariis incraffatis, inarticulatis, infernè denudatis, fupernè ramofiffimis, rubris; ramulis confertis, fub verticillatis, abbreviatis, multifidis, articulatis; articulis longitudine diametrum æquantibus. T. 85. & T. *G.*

Since I publifhed my defcription of this fpecies, it has been found on the fhores of Caithnefs and Orkney by Mr. Borrer and Mr. Hooker. Two kinds of fructification produced by this fpecies, from a drawing by Mr. Hooker, are reprefented in Plate G magnified 1.

149. *coccinea.* C. filis ramofiffimis, primariis incraffatis, hirfutis, inarticulatifque, coccineis; ramis alternatim decompofito-pinnatis; pinnulis ultimis multifido-fafciculatis, articulatis; articulis diametro fub brevioribus. T. 36. & T. *G.*

*β. tenuior.* Filis tenuioribus.

The variety *β* has been fent to Mr. Turner from the fouthern coafts by Mrs. Griffiths, and alfo from Ireland by Mr. Templeton and Mifs Hutchins. Its fize is more flender than that of *a*, and its ramuli fhorter, and lefs feathery. The moft remarkable difference however lies in the capfules, which inftead of being ovate, are lanceolate, and produce two rows of fmall globular feeds; they are feffile at the axillæ of the ramuli. The feeds of the ovate capfules, which in my T. 36 are reprefented globular, fhould, according to Mr. Turner's obfervations, have been made pyriform. The lanceolate capfules of the variety *B* are reprefented in my plate F, from a highly magnified drawing by Mr. Hooker.

b. *longitudinaliter venofæ.*

150. *elongata.* C. filis ramofiffimis, cartilagineis, craffis, reticulato-venofis, purpureis; ramis ramulifque elongatis, diffufis; articulis diametro dimidio brevioribus. T. 33.

Befides the fructification reprefented in T. 33; the minute lanceolate capfules alluded to in the defcription, are alfo reprefented in the fupplementary plate F, from a fketch by Mr. Hooker. Thefe pod-like proceffes, in which the fuppofed feeds are lodged, at length grow into branches.

151. *fufca.* C. filis ramofis, venofis, fufcis; ramis diftantibus, fub-alternis; ramulis patentibus clavatis; articulis medio tranfverfim fafciatis, diametro duplo longioribus. T. 95

152. *polymorpha.* C. filis dichotomis, venofis, faftigiatis, cartilagineis, atropurpurafcentibus; articulis centro punctatis, diametro fub-brevioribus. T. 44.

153.* *Brodiæi.* C. filis ramofiffimis, venofis, purpureo-nigrefcentibus; ramis elongatis; ramulis fparfis, patentibus, multifidis, fafciculatis; articulis ramorum obfoletis, ramulorum diametro fub-longioribus. T.

154. *fucoides.* C. filis ramofiffimis, venofis, diffufis, fubcartilagineis; fufconigris; ramulis horizontaliter patentibus, dichotomis, ultimis incurvis, acuminatifque; articulis diametro fub-fefquilongioribus. T. 75.

155. *nigrefcens.* C. filis ramofiffimis, venofis, ftrictis, fub cartilagineis, fufconigris; ramulis erectis dichotomis acuminatis; articulis diametro fub-fefquilongioribus.

C. *nigrefcens.* *Eng. Bot.* t. 1717. (exc. fyn.)

On the Beach at Yarmouth; *Mr. Turner.* Coaft of Devonfhire; *Mrs. Griffiths.* Brighton; *Mr. Borrer.*

I have been induced here to admit this fpecies under the name of C. *nigrefcens* in oppofition to the opinion of Sir Thomas Frankland, who has fent me the following plant by that name, becaufe I find that this is the plant fo called by moft Botanifts, and even as Mr. Turner affures me, by fome who were well acquainted with Hudfon. It fo ftrikingly refembles *C. fucoides* in the fize and color of the filaments, that it is not without fome hefitation that I publifh it as a feparate fpecies, but Mr. Turner who has repeatedly examined recent fpecimens of the two plants together, is decidedly of opinion that they are perfectly diftinct. It differs in having its main filament of far greater thicknefs than the reft, and the whole of its branches remarkably ftraight and erect, while the habit of the other is particularly bufhy. The outline too of the two fpecies is very diffimilar, that of *C. fucoides* being nearly orbicular, but that of *C. nigrefcens* narrowly cuneiform.

156.* *urceolata.* C. filis ramofiffimis, venofis, diffufis, rufo-fufcis; ramulis patentibus, brevibus; articulis caulis longis, ramulorum brevioribus. T. *G.*

*C. nigrefcens. Fl. Ang.* p. 602?

On Rocks and the larger Fuci in the Sea. On ftems of *F. digitatus*, and on Rocks oppofite the Bathing-houfe at Scarbro'; *Sir T. Frankland.* On ftems of F. digitatus in the Ifle of Wight; *Mr. Turner* and *Mr. Borrer.* Alfo on the fame Fucus on the Beach at Brighton; *Mr. Borrer.* Near Forres; *Mr. Brodie.* Devonfhire Coaft; *Mifs Hill.*

For fpecimens of this plant I have to exprefs my obligations to Sir Thomas Frankland, who, as mentioned under the preceding fpecies, fent it to me by the name of *C. nigrefcens* of Hudfon. Mr. Turner informs me that he has feen it in fome Herbaria marked by Mr. Lightfoot, " C. *urceolata, M. S.*" an appellation peculiarly appropriate, as the capfules differ in their fhape from thofe of every other Conferva, and approach, efpecially when dried, thofe of *Splachnum urceolatum* or *ampullaceum.* It moft commouly grows parafitically on the larger Fuci, and as remarked by Mifs Hill, looks then at firft fight like red wool. Its color in that ftate is a fine rich brown red, which would hardly be fuppofed from the dull black that it affumes in drying. The veins or tubes which compofe the filament are fewer than thofe of *C. fucoides* and bear more refemblance to thofe of C. *ftricta.* The joints towards the root are long, but become gradually fhorter as they approach the ultimate ramuli, in which their length fcarcely exceeds the diameter. The drawing at Plate G was made by Mr. Hooker from a ſpecimen which had been dried, and repreſents the end of a filament magnified 4, and alſo ſeparately the upper and lower joints magnified 1.

157.* *patens*. C. filis ramoſis, venoſis, ſub-diffuſis, roſeis; ramis ramuliſque ſparſis, patentiuſculis; articulis diametro ſub-duplo longioribus.

On Fucus digitatus, in the Sea, near Bantry; *Miſs Hutchins*. At Seaton, Devon; *Mrs. Griffiths*.

This ſpecies, for a ſpecimen of which I am indebted to Mr. Turner, is nearly allied to C. *ſtricta*, but the habit of the two plants is very different. It is of about the ſame ſize, but may be diſtinguiſhed by its more diffuſe growth, by its different ramification, and numerous ſhort lateral ramuli. The length of the joints in both ſpecies is ſubject to ſome variation, but thoſe of C. *patens* are comparatively ſhorter. Many of the ultimate ramuli in the ſpecimen now before me are ſwollen, and in theſe red globules may be obſerved, ſimilar to thoſe which in ſeveral of the other marine algæ are called ſeeds; but with all due deference to the opinion of my friend Mr. Turner, I muſt confeſs that I ſtill feel myſelf very doubtful of their real nature. The ſketch at Plate G was made from a dried ſpecimen by Mr. Hooker, and repreſents different parts of C. *patens* magnified with powers 5, 3, and 1 of his microſcope.

158. *ſtricta*. C. filis ramoſis, venoſis, ſtrictis, faſtigiatis, tenuibus, phœniciis; ramis dichotomis erectiuſculis; articulis diametro ſub-triplo longioribus. T. 40.

β. *diffuſa*. Filis diffuſis.

The plant which I have here arranged as a variety of C. *ſtricta*, was gathered in the neighbourhood of Bantry by Miſs Hutchins, and in her opinion is a diſtinct ſpecies. There is indeed at firſt ſight a ſtriking difference between them, but this gradually vaniſhes when the two plants are compared. It differs in its mode of growth, which is much more buſhy, and in its general outline, which is more orbicular. The common appearance of C. *ſtricta* retains its gloſſy red when dried, but the color of this variety then turns to a dull dirty brown. In their ſtructure when examined with a microſcope they however exactly agree, as well as in the fruit, which has been diſcovered in the former ſubſequently to its publication in this work, and conſiſts of ſmall ovate dark red capſules, ſeſſile, or nearly ſo on the upper branches.

159.* *fibrata.* C. filis ramoſis, venoſis, rubicundis; ramis dichotomis; ramulis ſub-faſciculatis, apice fibris pellucidis obſeſſis; articulis caulis longis, ramulorum diametrum longitudine æquantibus. T. *G.*

On Marine Algæ, near Forres; *Mr. Brodie.* At Cawſie, Murrayſhire; *Mr. Hooker* and *Mr. Borrer.*

The filaments, which grow to the length of about two inches, are branched with repeated dichotomies, and ſtrongly marked with longitudinal veins. Their ſummits are fringed with numerous, long, extremely ſlender, dichotomous, tranſparent fibres, of which from their extreme tenuity, it is almoſt impoſſible, eſpecially in a dried ſpecimen, to aſcertain the ſtructure, but they, I think undoubtedly are of the ſame nature with thoſe of C. *barbata.* Beſides an appearance of capſules in the dried ſpecimens now before me, I alſo obſerve ſeveral maſſes of looſe jelly, imbedding numerous pyriform ſeeds, and ſurrounded by a few ſhort ſegments reſembling an involucrum. I at firſt ſuppoſed that the fructification is nearly of the ſame nature with that of C. *corallina*, and that the appearance of capſules is occaſioned by the involucrum being compactly cloſed over the jelly to protect the yet unripe ſeeds, but Mr. Borrer ſays, "When I examined it freſh with Mr. Hooker, at Brodie, we ſaw the capſules as we thought them, ſplitting at the apex (I think into four ſegments) but it never ſtruck me that they were any thing analogous to the involucrum of C. *Corallina.*" In another ſpecimen now before me there are no capſules, but many of the joints are ſwollen, and each of theſe includes a dark colored globule, ſimilar to thoſe obſervable in many other Confervæ. At

Plate G; from a ſketch by Mr. Hooker, a branch of *C. fibrata* is repreſented magnified 3, and alſo the joints of the ſtem magnified 2.

160.* *denudata.* C. filis ramoſiſſimis, venoſis, diffuſis, fuſceſcentibus; ramis ſparſis, divaricatis, elongatis, remotis; articulis diametro ſub-ſeſquilongioribus. T. *G.*

In the Sea at Southampton; *Miſs Biddulph.*

Mr. Borrer favored me with ſpecimens of this ſpecies, which he received from Mr. Sowerby, but they are ſo imperfect at the apices, that without his aſſiſtance I ſhould not have ventured to publiſh it. The color is brown, and Mr. Borrer's largeſt ſpecimen is about four inches in length. The filaments are repeatedly branched: the branches iſſue almoſt at right angles and are placed without order, but uſually at conſiderable diſtances from each other, and Mr. Borrer in a letter ſays, "Mr. Sowerby told me that the points of all the ramuli were very long and ſlender when the ſpecimens were recent, and fell off when they were put into freſh water." The length of the joints is nearly the ſame throughout the plant, and is about half greater than the diameter. The capſules have not been diſcovered. I am indebted to Mr. Woods for the ſketch at Plate G, which repreſents the ramification of the natural ſize, and the joints when magnified 3.

161.* *badia.* C. filis ramoſis, venoſis, ſtrictis, rubro-nigreſcentibus; ramis elongatis; ramulis abbreviatis, remotis, ſub-ſimplicibus; articulis diametrum ſeſquilongioribus. T. *G.*

On the Beach at Haſtings; *W. Borrer, junr. Eſq.*

Mr. Borrer who has examined this plant whilſt recent, conſiders it as a diſtinct ſpecies, and ſo far as can be judged from a dried and ſomewhat imperfect ſpecimen, I entirely coincide with his opinion. He thinks that it is intermediate between C. *nigra* and C. *urceolata*, from both of which among other things it may be at once diſtinguiſhed by its joints, which are nearly of the ſame length in every part of the filament, and in that reſpect approaches more to *C. fucoides*. For the ſketch in Plate G, I am indebted to Mr. Woods; it reprefents a filament of the natural fize, and alfo the joints of the ftem and a ramulus magnified 3.

162. *nigra.* C. filis ramofis, venofis, rubro-nigrefcentibus; ramis elongatis; ramulis abbreviatis, remotis, multifidis, fub-penicilliformibus; articulis caulis longis, ramulorum triplo brevioribus.

C. *nigra.* *Fl. Ang.* p. 595. WITHERING. IV. p. 131.

C. *atro-rubefcens.* T. 70.

It is already mentioned in the Introduction that the fpecies which I publifhed with the name of *atro-rubefcens* is Hudfon's C. *nigra.*

163.* *fibrillofa.* C. filis ramofiffimis, venofis, rubris; ramis ramulifque fparfis, ultimis brevibus, multifidis, apicibus protenfis, fibrilliformibus; articulis inferioribus longis, fummis abbreviatis. T. *G.*

In the Sea. On the Beach at Brighton and Shoreham; *Mr. Borrer.* At Seaton; *Mrs. Griffiths.* Bantry Bay; *Mifs Hutchins.*

The neareft affinity of this fpecies is with C. *byffoides,* from which it may however be readily diftinguifbed by its more diffufe and irregular ramifications. The ultimate ramuli are tufted as in that fpecies, but they are lefs numerous, by far more flender, and more repeatedly dichotomous. Mr. Borrer who has attentively ftudied this plant whilft frefh from its place of growth, in which ftate alone thefe flender ultimate ramuli can be examined with much advantage, informs me, "that they are not compofed, like the other parts of the plant, of feveral parallel tubes, but are fimply tubular, and fpurioufly jointed *(utriculis matricalibus)*, the length of the joints many times exceeding the diameter." Mr. Borrer alfo favs, but which I have not myfelf obferved, that fimilar fibres occafionally occur in other fpecies of this fection which are ufually without them, and therefore queries whether they may not poffibly be a parafitical production. The capfules refemble thofe of C. *byffoides,* except that they are moftly raifed on fhort fruit ftalks. For the fketch of this fpecies at plate G, I am indebted to Mr. Hooker; it reprefents a branch magnified 4, and alfo the joints of the lower part of the filament magnified 3.

164. *byssoides.* C. filis decomposito-pinnatis, venosis, flaccidis, rubris; pinnis pinnulisque alternis, ultimis perbrevibus, multifidis, penicilliformibus; articulis inferioribus longis, summis abbreviatis. T. 58.

165.* *parasitica.* C. filis bipinnatis, venosis, rigidiusculis, fusco-rubris; pinnis pinnulisque alternis; articulis diametro sub-brevioribus.

C. *parasitica.* *Fl. Ang.* t. 604. WITH. IV. p. 142. *Eng. Bot.* t. 1429.

On Fuci. Coast of Yorkshire, Cornwall, and Dorsetshire; *Hudson.* At Scarboro'; *Sir T. Frankland, Bart.* Bantry Bay; *Miss Hutchins.*

166. *pennata.* C. filis pinnatis, venosis, rigidiusculis, olivaceis; pinnis sub-oppositis elongatis, approximatis, strictis, spinæformibus; articulis longitudine diametrum sub-æquantibus. T. 86.

Mr. Borrer has gathered at Beachy Head an unusually large variety of this species, with oblong pedicellated capsules.

167. *scoparia.* C. filis ramosis, venosis, rigidis, olivaceis; ramis alternis, sub-bipinnatis, confertis; pinnulis, brevibus, alternis, acuminatis; articulis longitudine diametrum sub-æquantibus. T. 52.

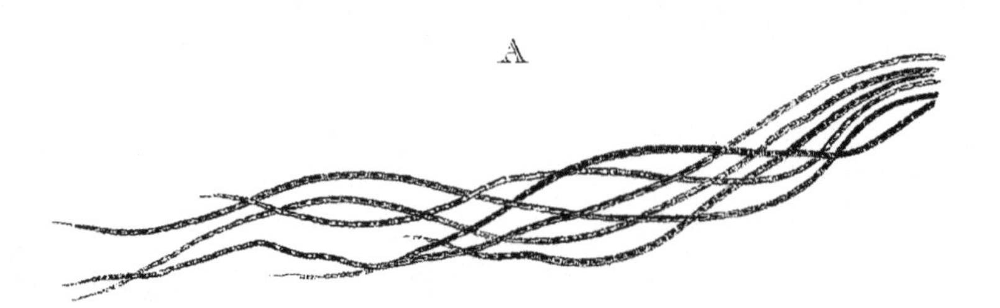


B

*Conferva ericetorum.*

Published by L.W. Dillwyn July 1 1802.

# CONFERVA ERICETORUM.

C. filamentis ſimplicibus tenuibus, denſiſſime implexis: diſſepimentis paulum contractis, articulis longiuſculis.

C. ericetorum. Roth Fl. Germanica III. p. 507. Cat. Bot. II. p. 206.

On moiſt Heaths about London and Yarmouth, &c.

---

THE learned and indefatigable Dr. Roth, of Vegeſack, near Bremen, was the firſt botaniſt who ever deſcribed this beautiful little Conferva, which he publiſhed in his valuable Flora Germanica, and Catalecta Botanica; two works to which I ſhall have frequent occaſion to refer in the courſe of the preſent undertaking.—My friend D. Turner added it to the Britiſh Flora, having found it growing abundantly on the bare parts of turfy heaths near Yarmouth, and compared it with ſpecimens ſent him by its firſt diſcoverer.

Its extremely ſlender ſimple filaments, of a dull purple colour, from half an inch to an inch in length, grow matted together in ſuch a manner that they form a denſe coat on the ſurface of the ground; and from their adhering ſo cloſely to it, as well as from the ſimilarity of their tint, are hardly diſtinguiſhable from the ſoil itſelf, except by one much in the habit of obſerving theſe plants. This is moſt probably the reaſon why it ſo long eſcaped notice, for it is common on all the moiſt heaths I have examined, and I cannot doubt its being equally abundant in ſimilar places throughout England. Some other ſpecies of Conferva delight in ſnch ſituations; but from theſe, the colour of the preſent plant is quite ſufficient to diſtinguiſh it. In Dr. Roth's figure above referred to, the interior ſubſtance is repreſented as having divided and collapſed towards each diſſepiment, whereas in all the ſpecimens which I have examined, the contrary has occured; and, as in many other Confervæ, it has formed an opake cylinder in the middle of each joint.

A. Filaments magnified 3.

B. Piece of ditto magnified 1.

B

B

D

# Conferva bipunctata.

Published by L. W. Dillwyn July 1 1802

F. Sansom sc.

# CONFERVA BIPUNCTATA.

C. filamentis ſimplicibus luteſcentibus lubricis, articulis brevibus cylindricis, bipunctatis.

C. bipunctata. Roth Cat. Bot. II. p. 204.

C. ſtellina. Muller in Nova Acta. Pet. III.

In Pools and Ditches; about London and Yarmouth, frequent.

---

THERE is reaſon to believe that this ſpecies, though not hitherto deſcribed by any Britiſh author, is ſufficiently common, particularly in the ſtagnant pools on heaths, either floating in thick maſſes on the ſurface, or looſe and ſtraggling at the bottom of the water. The firſt ſpecimens I received of it, gathered in Britain, were from my friend D. Turner; whoſe ſucceſs in his reſearches into almoſt every branch of Cryptogamia is too well known to need repetition here. Muller, who deſcribed and figured it as above quoted, ſeems to be the earlieſt author by whom it was noticed; though, from his work being incorporated in the tranſactions of the Peterſburg ſociety, the plant was but little known to botaniſts till publiſhed by Dr. Roth as a new ſpecies in the 2nd vol. of his Catalecta Botanica. I have adopted the name aſſigned to it by the latter botaniſt, not only on the ſcore of its ſuperior excellence, but alſo, becauſe the appellation given to it by Muller is apt to miſlead; being applicable only in a ſtate verging upon decay. The dots then aſſume a ſtellated appearance, as ſhown in the ſhorter filament of the figure A. in which the plant appears but ſlightly magnified. From C. ſpiralis it may generally be known by its larger ſize, more yellow and leſs gloſſy hue; from C. genuflexa I believe always by the former of theſe circumſtances, as well as by its being deſtitute of the broken appearance, which is a ſtriking characteriſtic of that plant. It is however difficult to diſtinguiſh theſe ſpecies with certainty, unaſſiſted by a microſcope; though with its aid, this may be immediately recognifed by the ſhortneſs of its joints, and by their containing each two dark ſpots, frequently furniſhed with a green longitudinal ſtreak running through them.— The form of theſe ſpots is in general almoſt elliptical, but ſometimes tends to globular; and, as above mentioned, they take in their latter ſtage a ſingular ſtellated appearance: the ſpace alſo that they occupy in the joints is far from certain, for ſometimes they fill nearly the whole, and at others only a ſmall portion of them.

Fig. D. repreſents what I ſuppoſe to be a variety of C. bipunctata, though it may poſſibly hereafter prove to be a diſtinct ſpecies; I found it abundant on Finchley Common, in March, 1802, in company with my friend J. Woods, jun. and both from its brown colour and the lingular formation of its ſpots, it differed remarkably from the general appearance of the plant.

A. Filaments magnified 4.
B. & C. Ditto in different ſtages 1.

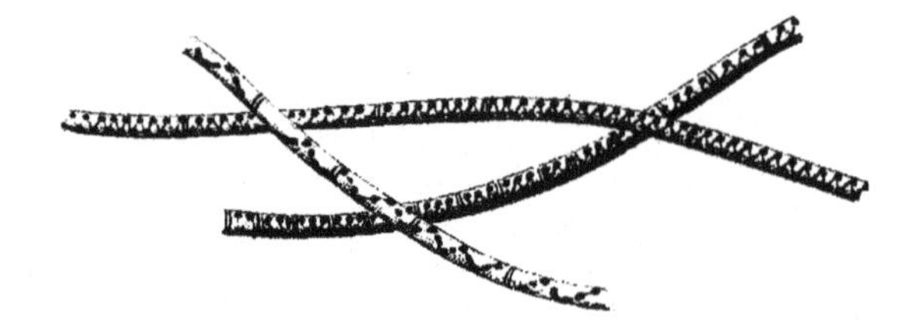

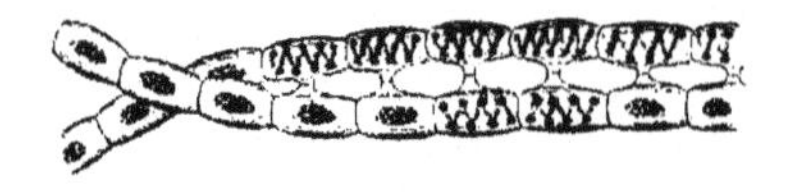

*Conferva spiralis.*

# CONFERVA SPIRALIS.

C. filamentis ſimplicibus lubricis articulis cylindricis longiuſculis, fructificationum granulis ſimpliciter ſpiralibus.

C. ſpiralis. Rot Cat. Bot. II. p. 202.

C. quinina. Muller in Acta. Nova. Pet. III.

In ſtagnant Ditches and Pools; about London and Yarmouth, common.

---

IT is not without conſiderable heſitation that I have ventured upon introducing this Conferva, as ſpecifically diſtinct from the following one, nor do I even now look upon the caſe as altogether certain, though I conſider that the regarding them as different, in compliance with the opinions of Muller and Dr. Roth, is the moſt likely way to avoid future confuſion.

C. ſpiralis is frequently found mixed with C. nitida and C. bipunctata; from which it is to be diſtinguiſhed by its much ſmaller ſize, and by the diſpoſition of its granules in a ſingle ſpiral tube, reſembling, as is obſerved by Muller, a chain of Roman V's. That botaniſt appears to have been the firſt who deſcribed it in the excellent paper above quoted; and of courſe I ſhould have adopted his name, which, though quaint, is very expreſſive, had it not been more generally known by the equally applicable one of Dr. Roth. Not only in its nature, but alſo in its colour, its mode of growth, and the places which it inhabits, the affinity between this plant and C. nitida is very great; as what is ſaid of the joints and granules of that ſpecies is equally applicable to this, I refer my reader to the remarks there given, and ſhall add nothing farther reſpecting C. ſpiralis, except a curious circumſtance mentioned in the Catalecta Botanica; which is, that if the water in which it is put be ſtrongly agitated, the granules looſe their ſpiral form and become ſcattered without order through the joint. I have however repeated this experiment without ſucceſs.

Since the defcription of C. fpiralis was written, and indeed the whole fafciculus finifhed, I have had an opportunity of tracing its growth fatisfactorily,* and of afcertaining that it is not C. nitida in a younger ftate; but was furprized to find that in the laft ftage of its exiftence, the filaments became connected in a manner precifely refembling C. jugalis, which ftrengthens the fufpicion that that curious plant is not a diftinct fpecies, but only an appearance affumed by C. nitida in certain fituations, or at certain periods of its growth; the fame circumftance will probably be found in fome other fpecies of this fingular tribe.

A. C. fpiralis magnified 1.

B. Ditto anaftomozing after the manner of C. jugalis, magnified 1.

* May 2d, I found C. fpiralis growing abundantly in a pool near Yarmouth, in which I obferved none when I examined it but a few days before; the filaments were then as reprefented in fig. 4.

May 6th, The plant occupied a larger fpace in the pool, but when magnified ftill appeared the fame.

May 10th, The plant was of a more dull colour and had loft fome of its lubricity, and when examined under a microfcope, many of the filaments were feen connected, as reprefented in fig. B. they differed from C. jugalis only in the difpofition of the feeds, being fingly fpiral in their fmaller fize, and in the oval maffes not appearing fo denfe in thofe joints wherein the granules had collapfed.

May 13th, The whole was in a ftate of decay, but all the joints which ftill retained the fpiral difpofition of the granules, had that difpofition only fingle; and though I examined a great number of filaments at each of the times above mentioned, I could not find one in which they were at all otherwife. This fudden appearance and difappearance of the Conferva had been before obferved by my friend D. Turner; who, in the Introduction to his Synopfis, p. 19, obferves, that often when he has known ditches filled with particular fpecies, he has returned after a fhort time and found not even a veftige of them left.

A

B

# CONFERVA NITIDA.

C. filamentis ſimplicibus ſplendenter lubricis, articulis longiuſculis cylindricis fructificationum granulis duplicato-ſpiralibus.

C. nitida. Fl. Dan. Tab. 819.

C. rivularis. B. Fl. Ang. 591. Fl. Scot. p. 976. With. IV. p. 128.

C. paluſtris ſenica, craſſior & varie extenſa. Dill. Muſc. 3. t. 2. f. 2.

Byſſus paluſtris confervoides non ramoſa viridis, filamentis craſſioribus, ſetas aprinas æmulantibus. Mich. Gen. p. 210. t. 89. f. 6.

C. decimina. Muller in Nova Acta. Pet. III.

C. ſetiformis. Roth Cat. Bot. Faſc. 1. p. 171. II. p. 203.

In Ditches and Pools; about London and Yarmouth, common.

---

THIS curious vegetable, which there is every reaſon to believe is not uncommon in ditches and ſtagnant waters throughout England, was near a century ago regarded as a diſtinct ſpecies by thoſe botaniſts, who at that time directed their attention to this tribe; though from their imperfect acquaintance with the ſubject, they reſted its claim to be conſidered as ſpecifically diſtinct from C. rivularis, only upon its ſhorter thicker filaments, and the ſtraggling mode of its growth: circumſtances which, as the accurate Dillenius obſerved, might be occaſioned by the diſſimilar places which the two plants inhabit. Subſequent writers regarded them merely as varieties, till the preſent was figured in the Peterſburg Tranſactions and the Flora Danica; and in the year 1797, Dr. Roth gave a complete account of it in the firſt volume of his Catalecta Botanica. It in general grows at the bottom of the water in looſe irregular patches, not ſufficiently matted to contain air bubbles, nor ſo much entangled as moſt of its congeners: its threads extend to a foot or more in length, and in thickneſs are about equal to the hair of the human head: its colour, when viewed in its place of growth, is ſo dark as often to appear almoſt black; but in this refpect is liable to confiderable variation. From C. rivularis it may at once be diftinguifhed, not only by its different mode of growth above noticed, but equally by its gloffy hue and far greater lubricity; from bipunctata and genuflexa by its darker colour; and from all thefe, by its curious internal ftructure; in which refpect however under the microfcope it approaches nearly to C. fpiralis, but differs in its larger fize, and in its granules not being difpofed in a fingle fpiral tube; to C. jugalis it is ftill more nearly allied, but has a lefs flaccid appearance to the naked eye, and is eafily diftinguifhed when magnified, by its want of connecting proceffes. It would be a fortunate circumftance for the arrangement of this tribe, if more dependance could be placed on the relative proportions of the length and thicknefs of the joints; but it frequently happens that the former is twice or thrice, or even more, greater in fome fpecimens than in others. This circumftance may account for the difference of the fpecific characters given to the prefent plant in the Flora Danica, and Catalecta Botanica; the former defcribing it 'articulis longis,' the latter, 'brevibus.' I have frequently feen filaments in the fame fpecimens that agreed with either; but have, confiftently with its moft general appearance, adopted a term between thefe two extremes. Muller's defcription, which is otherwife both curious and accurate, is on this account not always applicable; he fays that every joint contains four Roman X's, and thence derives the name that he has given it. The granules appear to be confined in fpiral tubes, and vary confiderably in fize as well as in the diftance of the tubes; being fometimes much crowded, and fometimes at a confiderable diftance from each other. In order to determine with more certainty than was otherwife poffible the nature of thefe granules, my friend D. Turner and I placed fome in a folar microfcope, and found them perfectly pellucid, of a homogenous nature, with no appearance of their being filled by any granular fubftance; which confirmed, in fome degree, an opinion before entertained, that thefe are not feed veffels, but the true fructification of the plant.

A. B. C. C. nitida. magnified 1.

A

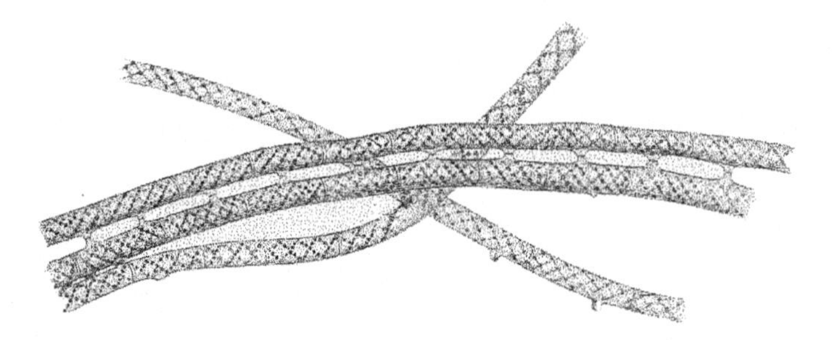

B

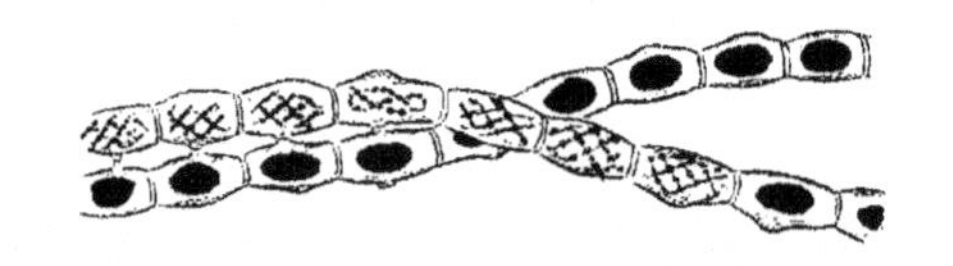

*Conferva jugalis.*

# CONFERVA JUGALIS.

C. filamentis ſimplicibus flaccidis, per paria ſœpe conjugatis, fructificationum granulis duplicato ſpiralibus, in globulos demum congeſtis.

C. jugalis. Flora Danica. Tab. 883.

C. ſcalaris. Roth, Cat. Bot. II. p. 196.

Pools and Ditches; near Yarmouth, Halſeworth, and other Places in Suffolk.

---

THIS plant, which in my opinion has a claim to be conſidered one of the moſt beautiful and intereſting of its tribe, was firſt made known to the botanical world by Muller, who gave a characteriſtic figure of it in the Flora Danica, as above quoted: it was afterwards found by Dr. Roth in the Dukedom of Bremen, and was laſt ſpring added to the Britiſh Flora by my friend Dawſon Turner, who detected it growing in ſhallow pools on Lound Heath, near Yarmouth; ſince which time we have together met with it in other places on the Northern part of Suffolk. What moſt ſtrikingly diſtinguiſhes it at firſt ſight is, its flaccid appearance rather reſembling that of the narrow varieties of Ulva compreſſa, and the ſeemingly great ſize of its filaments, ariſing from their coheſion; by which, and their mode of growth, which is looſely entangled, the naked eye may diſtinguiſh it from C. nitida, wherewith, when magnified, it has a ſingularly ſtrong reſemblance; ſo much ſo, that it may be doubted whether it is more than a variety of that plant: it agrees with it in ſize, in the general length of its joints, though I have not obſerved them ſo variable in C. jugalis as in that ſpecies, and in the ſpiral diſpoſition of its ſeeds; but differs in the latter collapſing from age into oval, or ſometimes globular maſſes, and alſo in the connecting proceſſes which form its moſt ſtriking character. Theſe are thrown out by many of the joints, and are extremely ſhort tubes, by means whereof moſt of the filaments attach themſelves to each other, and thereby receive a ladder-like appearance, whence Dr. Roth derived the excellent name of ſcalaris; which, however, as the plant was previouſly known by the equally applicable term of jugalis, I have declined adopting. In this reſpect the preſent ſpecies approaches the nature of C. genuflexa; but the yellower colour, ſmaller ſize, and broken appearance of that ſpecies, are ſufficient for the naked eye; and when magnified, its far different joints and mode of growth immediately diſtinguiſh it. Long filaments are often found wholly unconnected with the reſt, and ſometimes the uniting proceſſes iſſue only from one or two joints. I feel myſelf perfectly unqualified to offer the ſlighteſt conjecture on the purpoſes which the wiſdom of Providence has deſigned to anſwer by this ſingular union of the joints. Citizens Charles and Romain Coquebert, in a paper they communicated to the Philomatic Society of Paris, ſuppoſe that it is ſubſervient to the fructification, ſtating it to be "the firſt inſtance in the vegetable kingdom of a reproduction abſolutely analogous to that we find in animals;"* not only however may we obſerve granules in every reſpect ſimilar in thoſe joints which remain unconnected, but alſo in the filaments which ſometimes occur, in which not even the rudiment of a ſingle connecting proceſs is diſcernible. There is much curious matter concerning this Conferva recorded in Dr. Roth's Catalecta Botanica, which I cannot but regret that the limits of my preſent undertaking prevent my inſerting; I muſt therefore refer my readers to that work for farther information.

A. C. jugalis, magnified 1.

B. Ditto, in a more advanced ſtage, 1.

* Philoſophical Magazine, Vol. 3.

*Plate 6*

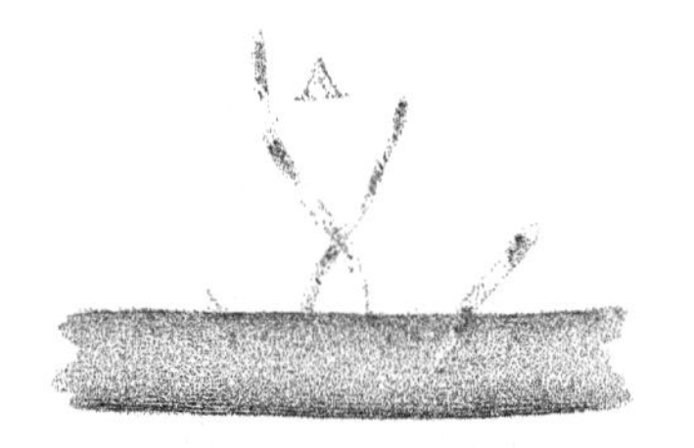


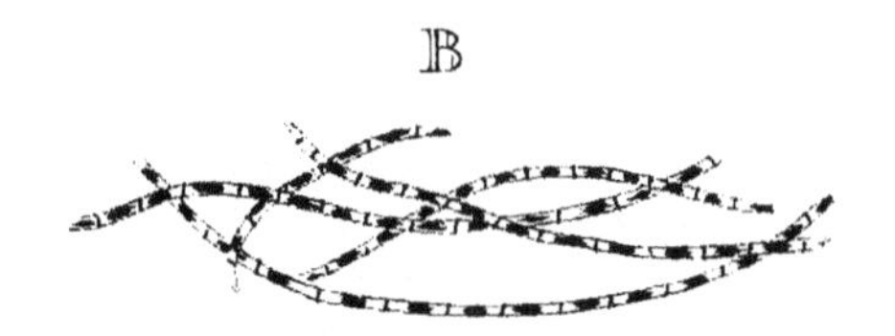


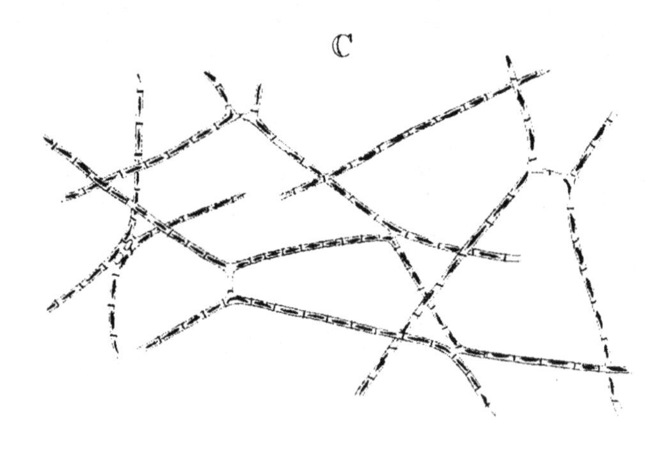


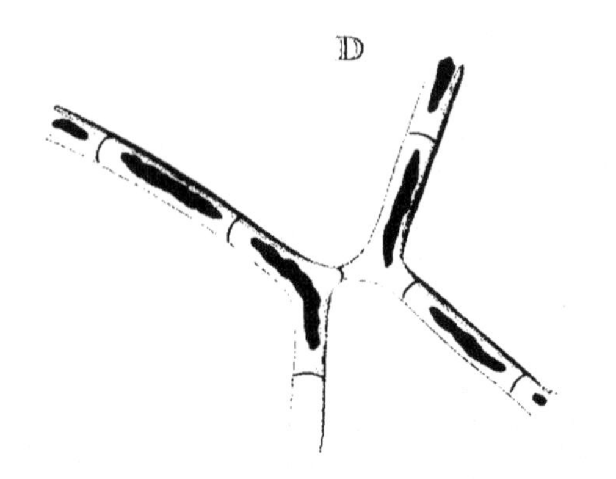

# CONFERVA GENUFLEXA.

C. filamentis ſimplicibus tenuiſſimis fragilibus hic illic genuflexis conjugatiſque; articulis longiuſculis cylindricis, granulis in lineas coacervatis.

C. genuflexa. Roth, Cat. Bot. II. p. 199.

C. ſerpentina. Muller in Nov. Act. Ac. Scient. Imp. Petrop. III.

In Ditches and Pools; about London and Yarmouth.

---

THE wonderful mode of growth, whence the preceeding ſpecies derives its name, is remarkable alſo (though in a far leſs degree) in the preſent, which is generally found floating in very thick maſſes on the ſurfaces of ditches and pools, and may be diſtinguiſhed by its ſhort filaments and pale yellow colour. When I firſt met with it in the vicinity of London, the threads were all ſimple, nor were there any ſymptoms of their having a tendency to anaſtomoſe, but their extreme brittleneſs ſeemed to be their moſt conſpicuous character, as all of them had the appearance of being more or leſs broken. Hence I concluded it to be the C. fragilis of Dr. Roth's Catalecta, (II. p. 204.) and I ſtill incline to this opinion; though, never having had an opportunity of examining any authentic ſpecimens, I have not ventured on quoting that as a ſynonym. It was not till the middle of April that I met with this ſpecies in the ditches about Yarmouth, and diſcovered it to be the real C. genuflexa, by comparing it with ſpecimens from Prof: Merteus, in the extenſive herbarium of my friend D. Turner. The length of the filaments does not appear to exceed an inch or two, though, from their brittleneſs, it is impoſſible to form an accurate judgment on the ſubject: this mode of anaſtomoſing is the ſame as has been already dwelt upon in the account of C. jugalis; but the connecting tubes are in general longer, and inſtead of iſſuing from almoſt every joint, they are placed at very uncertain diſtances, and the filaments are geniculate where they exiſt. C. genuflexa farther differs from jugalis in the threads not being regularly paired, but connecting themſelves with any other that is near them; in this reſpect, manifeſting a ſtrong affinity to C. reticulata.

In Plaiſtow marſhes I found a number of apparently ſeedling plants, of which I have added a ſketch, growing on C. rivularis; they ſeemed to adhere by a callus, which is probably the caſe with the conferva in general. Among theſe vegetables we muſt conſider the root as an organ of adheſion, not eſſential to the growth of the plant, as they continue to thrive when torn from it and floating on the ſurface of the water, nouriſhed probably by abſorbents, placed either in ſome particular part, or generally covering the frond.

| | | |
|---|---|---|
| A. | Seedlings of C. genuflexa growing on C. rivularis, magnified | 3. |
| B. | Filaments more advanced, | 3. |
| C. | Ditto anaſtomozing, | 3. |
| D. | A ſmall piece ditto, | 1. |

B

C

*Conferva muralis.*

Published by L. W. Dillwyn July 1. 1802.

# CONFERVA MURALIS.

C. filamentis fimplicibus tenuiffimis fafciatis rigidiufculis diffepimentis obfoletis; articulis breviffimis.

On moift Walls, Stones, Thatch, &c.

---

IT can hardly fail to ftrike even the moft cafual obferver of plants, that the green maffes obfervable on walls and ftones in damp fituations, muft owe their origin to vegetable matter. Ulva crifpa is known often to occafion them, but ftill more commonly do they proceed from the prefent plant; the minutenefs of which is fuch, that its having hitherto efcaped obfervation, is not wonderful, its filaments being fo fine that the human eye can fcarcely diftinguifh them; and it is only by the affiftance of the higheft powers of a compound microfcope that we can form any juft idea of their nature. Its mode of growth is very denfely matted, adhering clofely to the fubftance on which it grows, and infinuating itfelf into every crevice: it is compofed of threads about an inch long, equal at each extremity, varioufly twifted, and rather rigid; at leaft fo much fo, that when immerfed in water they do not follow the courfe of the current. Viewed with a good glafs, the filaments are feen to be compofed of extremely fhort joints, in general cylindrical, but fometimes affuming a globular appearance, interfected in an irregular manner by fafciæ: thefe I have obferved in fome others of this genus; they are feemingly of a diftinct nature from the diffepiments, being of a darker colour and thicker fubftance; but the moft remarkable difference confifts in their having nothing of that curved appearance difcernible in the others, which is occafioned by the cylindricity of the filaments. Some red ftriæ, doubtlefs of the fame nature, are reprefented in the figure of C. diftorta in the Flora Dan. Tab. 920, to which this fpecies bears a confiderable analogy; efpecially in the remarkably abrupt manner in which fome of the joints appear altogether colourlefs, leaving thofe with which they are immediately connected of their common green hue. I have frequently obſerved many ſmall grains attached to the filaments, but their minuteneſs is ſuch as renders it impoſſible to determine whether they are capſules, ſeeds, or only ſome extraneous matter.

A. Filaments magnified, 3.
B. Small piece ditto, 2.
C. Ditto ditto, 1.

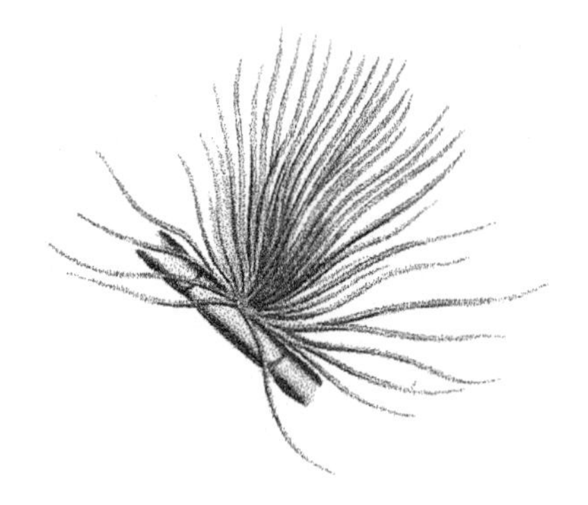

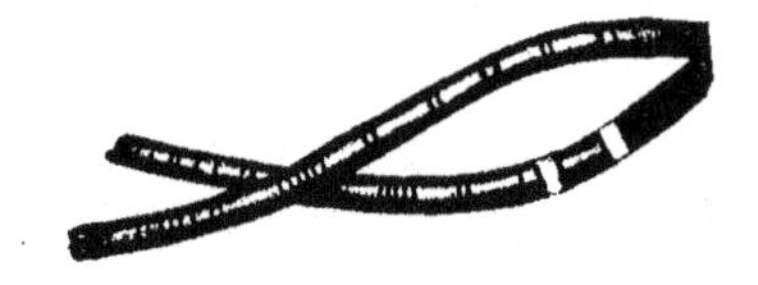

*Conferva confervicola.*

# CONFERVA CONFERVICOLA.

C. filamentis ſimplicibus minutis, ſub confertis acutis; diſſepimentis obſcuris, articulis cylindricis longitudine inequalibus.

C. marina paraſitica, tenuiſſima & breviſſima glauca. Dill Muſe. p. 552, t. 85, f. 21.

In the Sea, adhering to Fuci & Confervæ.

---

THIS delicate paraſite is by no means unfrequently found, in the lateſt months of autumn, on Fucus purpuraſcens, ſubfuſcus, Conferva elongata, rupeſtris, and other Confervæ; attaching itſelf principally to the ends of the branches, and often entirely covering them. It may be readily diſtinguiſhed by its very ſhort ſimple ſlender filaments, rarely exceeding one-eighth of an inch in length, and their dark glaucous colour. As well as in ſome other of the ſmaller unbranched ſpecies of this genus, the diſſepiments are not placed in any regular order, but at various diſtances from each other; and among them faſciæ frequently appear, nearly ſimilar to thoſe deſcribed under the laſt ſpecies.

There can be no doubt of this being really the plant deſigned by Dillenius, in the place above quoted, and called by him 'Conferva upon Conferva,' though Dr. Roth, in the firſt volume of his Catalecta, has referred that ſynonym to his C. mucor, which ſeems to be a different plant; and if we may judge from his account of it, may probably be ſome not uncommon paraſitic ſpecies in decay.

A. C. confervicola natural ſize, growing on Fucus purpuraſcens.

B. Ditto, on C. rupeſtris, magnified 3.

C. Small piece ditto, 1.

A

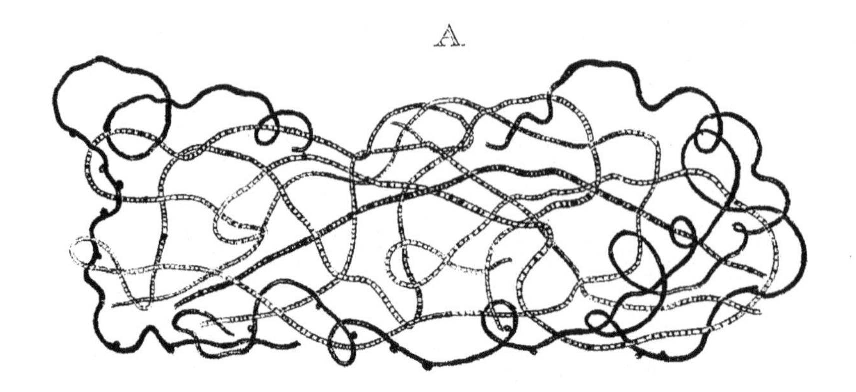

B

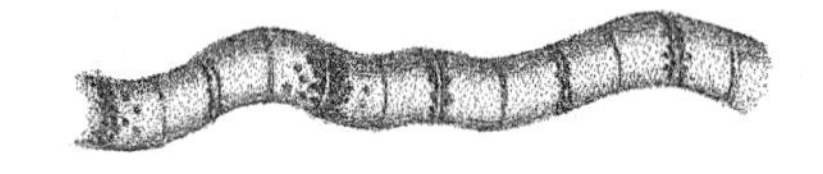

C

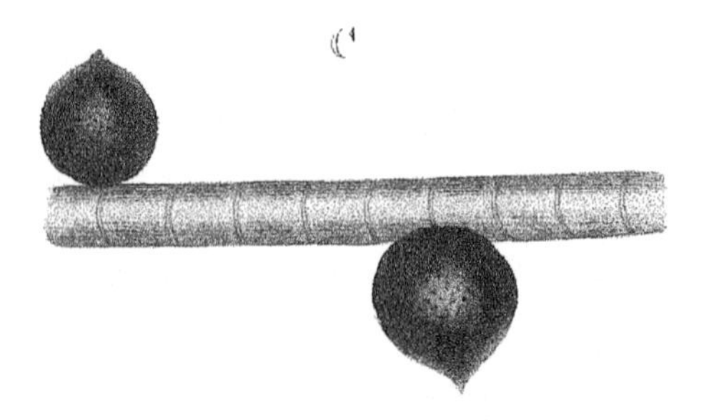

# CONFERVA CAPILLARIS.

C. filamentis ſimplicibus teretibus rigidiuſculis criſpatis implicatis fragilibus; diſſepimentis pellucidis; articulis cylindricis brevibus; capſulis ſeſſilibus.

C. capillaris. Sp. Pl. p. 1636. Fl. Ang. p. 598. Fl. Scot. p. 988. With. IV. p. 135.

C. Linum. Fl. Dan. t. 771. Roth. Cat. Bot. I. p. 174.

C. filamentis longis geniculatis ſimplicibus. Dill. Muſc. p. 25. t. 5. f. 25. A.

C. paluſtris, ſive Filum marinum anglicum. Ray. Syn. p. 60. n. 16.

C. geniculata minima noſtras. Moriſon. Hiſt. Ox. p. 644. ſ. 15. t. 4. f. 4.

In the Ditches and ſtagnant Pools of Salt Marſhes.

---

THIS ſpecies, no uncommon inhabitant of ditches near the ſea, may, at firſt ſight, be diſtinguiſhed from all others by the thickneſs of its filaments, which in ſize are equal to large thread; by their brittle and rigid nature when freſh; by their never adhering together, and by the remarkably curled and entangled mode of its growth; from which circumſtance I have hitherto found it impoſſible to trace with ſatisfaction either the root or apex of the plant, each end having an equally truncated appearance. The filaments extend to the ſurpriſing length of three or four feet; their colour is a pale yellowiſh green; the diſſepiments are quite pellucid, but unleſs carefully examined they appear darker than the joints, there being a thin blackiſh line on either ſide of them: in many filaments they are extremely apparent to the naked eye, and ſome of them, even without the aſſiſtance of a glaſs, may be ſeen to be ſwelled and much blacker than the reſt, which, in a ſpecimen now before me, is the caſe in every fourth, but in ſome others I have not found them ſo regular; this dark appearance, when highly magnified, proves to be occasioned by sub-elliptical granules imbedded in the filament, as is represented in the figure B. Some few are also found scattered in other parts of the joints, and I never doubted that these formed the fructification, till on the 8th of May, 1802, to our great satisfaction, my friend D. Turner and myself found the plant in the ditches about Yarmouth, copiously producing sessile roundish pointed capsules, precisely resembling those of C. dichotoma, &c.

C. capillaris, after it has been but for a few minutes exposed to the air, becomes perfectly flaccid, and when dried, the joints assume a kind of irregular alternately compressed appearance, which induced Linnæus, who evidently had seen only specimens in that state, so to describe it; but though he has in this instance been copied by Hudson, Lightfoot, and many other authors, this appearance is by no means so constant as to justify the stress he has laid upon it; and hence Dr. Roth, who found it apply better to the plant represented by Dillenius, t. 25. f. 5. B. which he believes to be specifically different, applied the appellation *capillaris* to that, and made the present a new species, under the name of C. linum. From the references nevertheless in the Species plantarum, I have very little doubt of ours being in reality what is there intended. I have subjoined a mark of nucertainty to Morison's figure, because he has drawn it as if it grew in the manner of a Chara.

This species is sometimes found in the pools near Yarmouth, rolled up into balls by the action of the waves, so as to resemble C. ægagropila. It differs from most others in not adhering to glass or paper after it is dried; nor does it, when once it has from that cause suffered contraction, ever recover its natural form by subsequent immersion.

A. C. capillaris of its natural size.

B. Ditto, without capsules, magnified 2.

C. Ditto, with capsules, magnified 2.

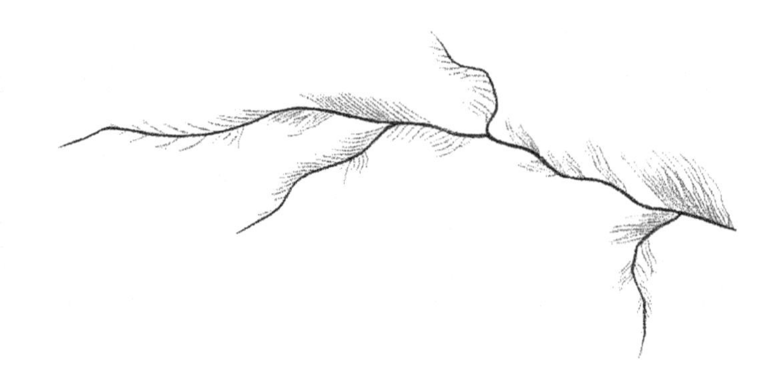

B

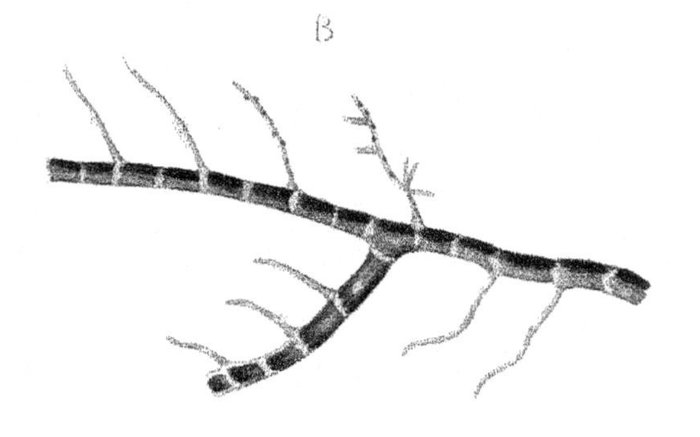

*Conferva flexuosa*

# CONFERVA FLEXUOSA.

C. filamentis dichotomis rigidiufculis; ramis flexuofis; ramulis fubfimplicibus tenuiffimis, alternatim fecundis patentibus, articulis cylindricis, diffepimentis obfoletis.

C. flexuofa. Fl. Dan. Tab. 882.

In the Pools in Yarmouth Salt Marfhes.

---

THIS beautiful fpecies was added to our Britifh Flora by D. Turner, Efq. who firft gathered it from among the rejectamenta of the fea at Yarmouth, and afterwards difcovered it growing abundantly near that town; it has not hitherto been found in any other part of England; but thefe plants have been fo little attended to, that it may poffibly not be uncommon in fimilar fituations in this ifland.—— The filaments grow in clofely entangled maffes at the bottom of the pools, and extend to the length of from four to eight inches; they are finer than the hair of the human head; their mode of growth is remarkably flexuofe; their fubftance rather rigid, and deftitute of all lubricity; their colour in general a pale yellowifh green about the apices, but fo dark as to be almoft black in the main fhoots; they are feldom more than once or twice divided, but are from bafe to fummit befet with fpreading fimple ramuli, often half an inch long, alternately arranged on each fide of the fhoots, and fo fine at their extremities as to be almoft invifible. No appearance of diffepiments can be detected without the ufe of the microfcope, and even then they are very faint, and of a paler colour than the intervening joints.— Some opake oval granules are frequently found fcattered on the branches, and the ramuli appear to be filled with others which are lefs in fize and more pellucid; but whether either of them are the fructification I cannot pofitively decide, though I fufpect it is not the former, having found exactly fimilar ones attached to other fpecies. On this, as well as on many different Confervæ, fmall pellucid tubes, which I have reprefented on one of the branches, and which are fuppofed by Dr.

Roth to be a ſpecies of Polypus, may frequently be ſeen adhering in an irregular manner to its ſurface. The Conferva introduced by Dr. Roth in the ſecond faſciculus of his Catalecta under this name, is a very different plant, and appears to be only a ſmall variety of Ulva compreſſa.

A. C. flexuoſa, of its natural ſize.

B. Ditto, magnified 3.

B

C

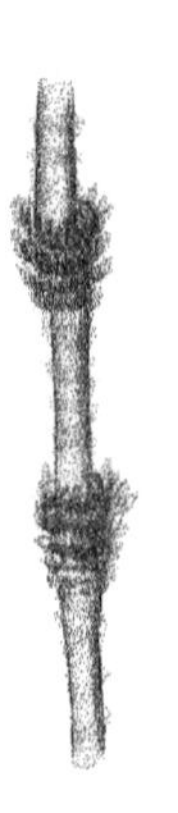

# CONFERVA ATRA.

C. filamentis ramofiffimis moniliformibus fub-gelatinofis; ramulis fetaceis, articulis apicem verfus dilatatis ciliatis, ciliis verticillatis imbricatis.

C. atra. Fl. Ang. p. 947. With. IV. p. 134. Eng. Bot. t. 690.

C. fontana, nodofa, lubrica, filamentis tenuiffimis nigris. Dill. Mufc. p. 39. t. 7. f. 46.

In Rivulets and Springs; in a fmall Rivulet flowing into *Gors Velen* Lake, near *Llanfaethly*, and in a Spring called *Ffynnon bach y Lusg* in *Gors Bach*, between the Church and *Trefadog*, in the Ifle of Anglefea, *Brewer*. Near Martin in Surrey, *Hudson*. Near Croydon, *Dickson*. At Lound, near Yarmouth, *D. Turner, Esq.*

---

THIS rare and beautiful fpecies, though well reprefented by Dillenius, and defcribed by Hudfon, appears to have been but imperfectly known to modern Botanifts, till it was figured in Englifh Botany from a fpecimen gathered in a rivulet at Hopton near Yarmouth, by my friend Dawfon Turner, who favoured me with the magnified drawing, fig. 2, from the delicate pencil of Mrs. Turner. It deferves, perhaps, to be confidered one of the moft rare of the Britifh fpecies; its nature and appearance being fuch, that it is hardly poffible to fuppofe it fhould have been often overlooked, nor does it feem to be known to Dr. Roth, or any foreign botanift. The places of growth that it prefers, are limped rivulets, where it is found mixed with C. gelatinofa, to which it has more affinity than to any other known plant of the genus. Its colour, which in its earlieft ftage is a pale green, varies in the feveral periods of its growth through the different fhades of green, till at laft it becomes almoft black: on a clofe examination to the naked eye, it has the appearance of a ftring of minute beads, which when the plant is highly magnified, proves to be occafioned by each joint being thickened towards its apex by whorls of very minute mealy fibres, having ſomething like a jointed appearance, but ſo minute as to make it almoſt impoſſible accurately to determine their nature.—There is no danger of its being confounded with any other ſpecies.

A. C. atra, of its natural ſize.

B. Ditto, magnified 3.

C. Ditto, magnified 1.

A

B

# CONFERVA MUTABILIS.

C. filamentis ramosissimis gelatinosis sub-moniliformibus; ramulis penicilliformibus, setaceis, ramosis; dissepimentis contractis, articulis brevibus.

C. mutabilis. Roth Cat. Bot. I. p. 197.

C. gelatinosa, γ. Fl. Ang. p. 598. With. IV. p. 135.

Conferva 'espece non decrite.' Vaucher in Journal de Physique, LII. t. 3. f. 7.

C. stagnalis, globulis virescentibus mucosis. Dill. Musc. p. 38. t. 7. f. 44.

In Ditches and Rivulets adhering to Sticks, Stones, or decaying Vegetables; about London and Yarmouth, common.

---

IT is sufficiently known that five of the Confervæ nodosæ, which Dillenius in his Hist. Musc. described as distinct, were afterwards united by Linnæus in the Spec. Plant. into a single species, under the name of C. gelatinosa. Of these, the first, second, and fourth, though submitted to a microscope, exhibit no farther difference than that of colour. The fifth is C. atra, figured in the preceeding plate; and the third, which is the plant now before me, even if but slightly magnified, instead of the short crowded verticillated ramuli, which occasion the characteristic bead-like appearance of C. gelatinosa, arrests the attention of the observer by its delicate pellucid almost colourless shoots, beset on each side by a number of very minute green tufts of ramuli, disposed generally in opposite directions. Dr. Roth seems to have been the earliest among modern botanists who accurately ascertained its nature, and he published it with a well-defined character in the first fasciculus of his Catalecta Botanica, giving to it the name of mutabilis, on account of a very singular change that he observed it undergo, in different periods of its existence.— He has erred in referring, as a synonym to Dillenius, 'C. fluviatilis sericea tenuis,' t. 6. f. 34; but his mistake is by no means surprising; for that figure not badly expresses the general habit of the plant; and the etching above referred to, which from Dillenius's Hebarium, and original drawings in the collection of Sir Joseph Banks, is known to have been designed for C. mutabilis, is very coarsely executed. It was first pointed out to me as a distinct species by my friend D. Turner, who gathered it near Yarmouth; and I have since found it in considerable abundance in most of the pools and ditches about London. Its length varies from half an inch to three inches, and its colour from a light to a dark green. The main shoots are nearly colourless, and formed of numerous short joints, contracted towards each end, and containing in their middle a band of granules, which we must suppose to be the fructification of the plant; though from its near affinity to C. gelatinosa one would be rather disposed to look for the same sort of fruit as is found in that species. At their dissepiments, the stems throw out small tufts of green ramuli, scarcely equal to one-fourth of their thickness, and so divided and sub-divided into extremely minute expanding branches, as to give them a pencil-like appearance; in some specimens they are of a compact oblong form, and in others more lanceolate, with the extreme branches considerably lengthened out.

For the drawing I am indebted to my friend Joseph Woods, jun. F. L. S.—— B represents a piece, which, though not so beautiful as many that might have been selected, we thought better calculated to give a clear idea of the plant.

A. C. mutabilis, natural size.
B. Ditto, magnified 1.

B

*Conferva glomerata*

Published by L. W. Dillwyn Nov 1 1802

F. Sansom sc.

# CONFERVA GLOMERATA.

C. filamentis ramofiffimis; ramis alternis; ramulis fecundis, fafciculatis, penicilliformibus; diffepimentis pellucidis; articulis cylindricis longiufculis.

C. glomerata. S. Plant. p. 1637. Fl. Ang. p. 692. Fl. Scot. p. 993. With IV. p. 140. Fl. Dan. t. 651. 2.

C. criftata. Roth. Cat. Bot. I. p. 193. II. p. 220. Fl. Germ. III. p. 512.

C. fontinalis ramofiffima glomeratim congefta. Dill. Mufc. p. 28. t. 5. f. 31. Ray. Syn. p. 59. n. 8.

C. viridis capillacea, brevioribus fetis, ramofior f. Conferva minor ramofa. Morifon. Hift. Ox. III. p. 644. f. 15. t. 4. f. 2.

On Stones and Wood in clear Rivers and Streams.

---

THIS elegant fpecies delights in the pureft waters, and, as may be concluded from its appearing in nearly every Flora, adorns moft of the limpid ftreams in Europe.—The root is a fmall callus, whence arifes the principal ftem, varying from two or three inches to a foot in length, and repeatedly divided and fubdivided; the ultimate branches are alternate, and befet on the upper fide with a ramulus at the end of nearly every joint, fo as to give them a bufh-like appearance, which is highly characteriftic of the plant. The fructification has not yet been difcovered, but I think there can be little doubt of its confifting in capfules nearly fimilar to thofe figured in the other fpecies. Linneus has erred in the fynonyms of this plant in the Species Plantarum, he having there referred to 'Dill. Mufe. 28. t. 5. f. 32. and t. 5. f. 28 & 29.' though under C. vagabunda which immediately precedes it, he had before referred to t. 5. f. 32, and again to t. 5. f. 29, as his C. rupeftris. He alfo fortunately gave as a fynonym of this plant Morifon's C. viridis capillacea, &c. to which Dillenius refers, as his C. fontinalis ramofiffima glomeratim congefta, t. 5. fig. 31. and which, with the former, are good repre- fentations of this plant. Dr. Roth has united C. glomerata and C. sericea into one fpecies under the name of C. criftata, and I have but little doubt that he has confounded two fpecies, as an authentic fpecimen with which I was favored by my friend Dawfon Turner, is certainly diftinct from glomerata, which he muft have included in his defcription, by his referring to the excellent figure in the Flora Danica.

Though liable to confiderable variations, as well in the length and thicknefs of the filaments, as in their being fometimes more or lefs branched, yet it may be at firft fight diftinguifhed from other fpecies, by its beautiful green color, and cha- racteriftic bufh-like ramuli. It is often much infefted with C. flocculofa, which grows parafitically on it, and fometimes nearly covers it. It adheres to both glafs and paper.

For the drawing I am indebted to my friend Jofeph Woods, jun. F. L. S.— it well reprefents the plant in a rather advanced ftage of growth.

A. C. glomerata, natural fize.
B. Ditto, magnified 3.

*Plate 14*

# CONFERVA FRACTA.

C. filamentis ramofiffimis implexis; ramis ramulisque divaricatis; articulis adultioribus oblongis junioribus cylindricis; capfulis feffilibus fub-rotundis.

C. fracta. Fl. Dan. T. 946.

C. divaricata. Roth. Cat. Bot. I. p. 179. t. 3. f. 1. Fl. Germ. III. p. 510.

In ftagnant Ditches and Pools.

---

THIS fpecies was firft defcribed and figured in the Flora Danica under the name of C. fracta, and afterwards by Dr. Roth in his Catalecta Botanica, who was not then aware of its having been pre-defcribed, under the name of C. divaricata: the former name appears to be moft eligible, not only on account of its priority, but alfo becaufe it is peculiarly characteriftic. I firft detected it near Yarmouth; afterwards copioufly producing feffile capfules in Lock fields, near London, and fince in many other places, and I think there is little doubt of its being one of our moft common fpecies. It grows in denfely entangled maffes, generally floating on the furface of ftagnant waters, and is of a dull dark green color. The filaments vary in length from one to four inches, are equal in thicknefs to the human hair, rather rigid, and divided and fubdivided into branches in an irregular manner: the branches are divaricate, moft commonly alternate, but fometimes feveral together are difpofed on the fame fide: in length they differ very much, fome being long, and others fo fhort, and apparently abruptly terminated, as to give the plant a broken appearance, which is highly characteriftic, and by which, and its divaricate ramifications, it may be diftinguifhed from its congeners. The joints, which otherwife are cylindrical, frequently appear to be fwelled, and affume an oblong form. This appearance I have alfo obferved, though far lefs frequently, in C. littoralis, rofea, and fome others, and I fuppofe that it muſt be attributed to age. It is often much infeſted by polypi. Profeſſor Mertens is of opinion that this is the plant intended by Dillenius, in his Hiſt. Muſc. t. 3. f. 11. and named C. bulloſa by ſubſequent authors, but the ſpecimen in Dillenius's Herbarum is certainly another ſpecies, and I feel no heſitation in adopting D. Turner's opinion, that many of thoſe plants which grow ſufficiently entangled together to retain bubbles of air, and are thereby floated on the ſurface of the water, have been confounded together by all authors under that name, and conſequently that the Confervæ bulloſæ are a family, and not a ſpecies of this tribe.

After being dried, the Confervæ bulloſæ have been uſed as wadding for ſtuffing garments, and wove into coarſe houſehold linen. Weis in his Plantæ Cryptogamicæ Floræ Gottingenſis, page 23, relates that formerly the river Unſtrut, after inundating a large tract of country in Upper Saxony, on again retiring into its proper channel, left a great quantity of C. bulloſa, which, having been gathered and dried by the inhabitants, was uſed by them for ſtuffing their garments, but that it occaſioned violent pains in their limbs. It is alſo uſed for making coarſe paper.

A. C. fracta, magnified 3.
B. Ditto 1.

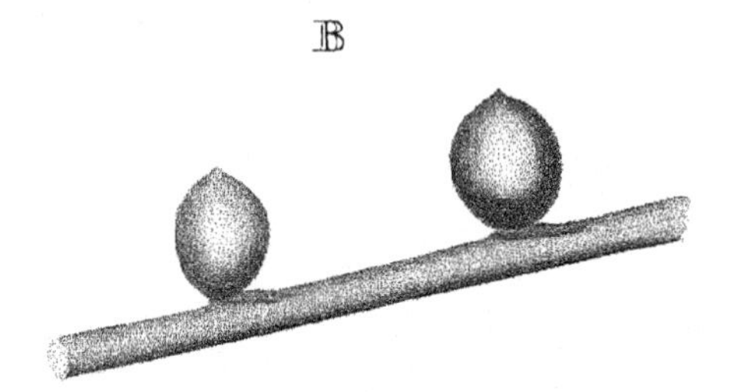


*Conferva dichotoma*

# CONFERVA DICHOTOMA.

C. filamentis, fafciculatis ftrictis faftigiatis dichotomis fub-articulatis, diffepimentis obfoletis; articulis longiffimis, capfulis ellipticis feffilibus.

C. dichotoma. Sp. Pl. p. 1635. Fl. Ang. p. 593. Withering. IV. p. 129. Eng. Bot. t. 932.

Ceramium dichotomam. Roth. Cat. Bot. I. p. 153. Fl. Germ. III. p. 474.

C. dichotoma fetis porcinis fimilis. Dill. Mufe. p. 17. t. 3. f. 9.

C. Plinii, fetis porcinis fimilis. Ray. Syn. p. 58.

In Ditches, common.

---

THIS Conferva, in denfe maffes, occupies, and often nearly fills the ditches in many parts of England, throughout the fpring and fummer months; confpicuous for its dark green color, matted appearance, and above all, its erect faftigiated fummits, which, at firft fight, bear a ftrong refemblance to a parcel of hog's briftles, to which they are aptly compared by Dillenius. The filaments are membranaceous, tubular, filiform, in general about two feet long, and confiderably thicker than horfe hair, always ftraight and fimple, or but once or twice divided, till they arrive at a few inches from their apices, when they are branched with repeated dichotomies, at uncertain but fhort diftances from each other, the angles of the divifions being every where acute. The fummits are blunt; the length of the joints irregular, though always confiderable; in a frefh ftate, their beginning and termination can hardly be difcovered, but, after the plant is dried, they appear flightly contracted at each end. The capfules, which were firft difcovered by Profeffor Mertens, of Bremen, are rather longer than the width of the filaments, and fcattered without order about them, fometimes fingly, and fometimes in clufters of five or fix together. A doubt is fuggefted in Englifh Botany, whether thefe are true capfules, or only fome extraneous bodies; I have however been enabled to decide that it is without foundation, and that the feeds efcape, as I believe they do univerfally in thofe plants, which conftitute the genus ceramium of Roth through an aperture, which, when the feeds are matured, is formed at their apices. They are found only in the fpring.

C. dichotoma grows about the bottom of ditches; as it approaches decay, it rifes to the top of the water, and there expofed to the fun and air lofes not only its natural form, but alfo its color, turning to a pale yellowifh green, and, becoming inflated with air, bubbles like many other frefh water fpecies. Few plants of this tribe have been either longer or better known. When dried, it becomes rigid, and adheres but very flightly to either glafs or paper.

A. C. dichotoma, natural fize.

B. Ditto magnified 2.

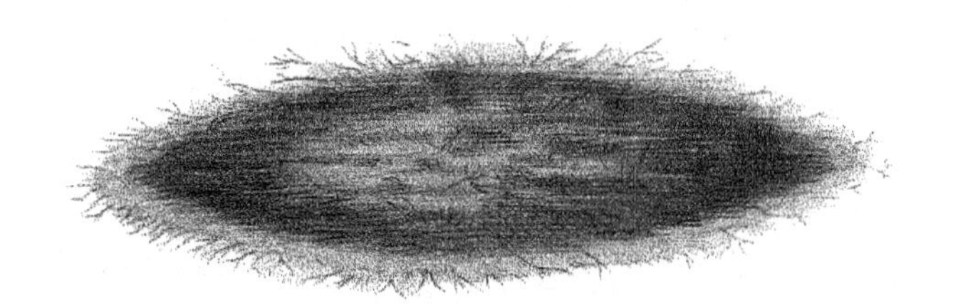

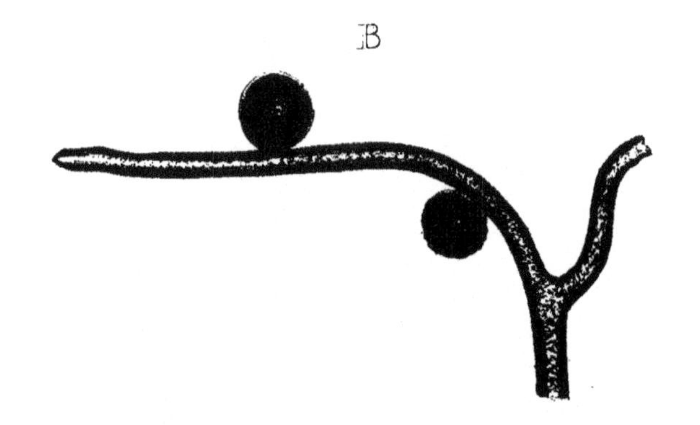


*Conferva frigida*

# CONFERVA FRIGIDA.

C. filamentis inarticulatis repentibus ramofis; ramis fubdichotomis alternis, exficcatione diftinctis; capfulis feffilibus rotundis.

C. frigida. Roth. Cat. Bot. I. p. 166. Fl. Germ. III. p. 491.

C. amphibia fibrillofa et fpongiofa. Dill. Hift. Mufc. p. 22. Tab. 4. f. 17. A.

C. terreftris exilis fibrillofa. Ray. Syn. p. 59. n. 7.

On the Ground in moift fhady places.

---

THIS Conferva, not unfrequently found in turnip-fields during the winter and early months of fpring, particularly in a northern expofure, and cold foil, had been confidered by Dillenius and all other writers, as not fpecifically diftinct from C. amphibia, till Dr. Roth feparated it in the firft volume of his Catalecta Botanica. His reafons for thus doing appear to me fo convincing, that, though in all matters of this nature, I would wifh to proceed with the utmoft caution, I have felt no repugnance in adopting them, and am convinced my reader will not be difpleafed at my introducing them at foot in the words of their author.* Dillenius, as well from his figure as defcription evidently knew both fpecies; though, not accuftoming himfelf to the ufe of a microfcope, he regarded them as the fame. How far Hudfon, Lightfoot, and Withering were equally acquainted with both, may perhaps admit of fome doubt: for my own part I fhould be inclined to think that they refer to this alone, but their defcriptions are of fuch a

* "Habitum enim fuum fub quavis conditione retinet et rami vel ramuli diftincti exficcatione nunquam coëunt in apices rigidiufculos, aculeos referentes, nec unquam in altum excrefcunt, fed depreffi repent. Septentrionalem regionem et frigidam femper fpectat illamque tantum amat, madifactione fpongiæ in modum aquam non imbibit, ut *Conferva amphibia* nec in majus volumen fefe expandit. Differt infuper. 1. ftratis filamentorum tenuioribus, laxe et inordinate expanfis et depreffis: nec craffis, fuperiore fuperficie quafi reticulatis et filamentorum ramulorumque apicibus erectiufculis, 2. Filamentis rigidioribus, quafi herbaceis, obfcuris, linea tantum longitudinali pellucida praeditis, cum ftriis tranfverfalibus in cortice: nec membrana tenuiffima fub pellucida, tubulofa, finuosa et rugofa abfque corticis evidentioris, quafi herbacei, veftigio compofitis." Roth. Cat. Bot. I. p. 168.

nature that it is a matter of very little confequence. C. frigida covers the ground generally in irregular patches two or three inches in diameter, of a rather pale green color, very flightly adhering to the foil, and if examined while growing, is feen to form feveral ftrata of loofe unconnected filaments. Its mode of ramification is not altogether dichotomous, but it rather feems to throw out a feries of alternate branches iffuing at acute angles with the ftem. The filaments are hardly fo large as human hair; their length is probably about an inch, but this, from their matted mode of growth, cannot certainly be detected; they are very flaccid, and when the plant is taken up, fall together, but are wholly deftitute of lubricity, fo that after they are dried, they neither adhere to paper or glafs. In this ftate they turn to a pale yellowifh green. The capfules which D. Turner and myfelf firft found in a field adjoining the ruins of Burgh Caftle, muft be confidered very rare, from their having efcaped the notice of Dr. Roth, and his indefatigable friend Profeffor Mertens; they are but thinly fcattered over the filaments. Even under the higheft powers of a microfcope the frond exhibits no appearance of any tendency to articulation. Dr. Roth, in the fecond volume of his Catalecta, page 217, defcribes a fpecies under the name of C. arenaria, which, at firft fight, he fays may be taken for C. frigida. I have not at prefent feen this, though I hope hereafter to be able to add it to the Britifh catalogue, having no doubt but that many more Confervæ will be found growing on the ground which have at prefent efcaped our notice.

A. C. frigida, natural fize.

B. Ditto magnified 1.

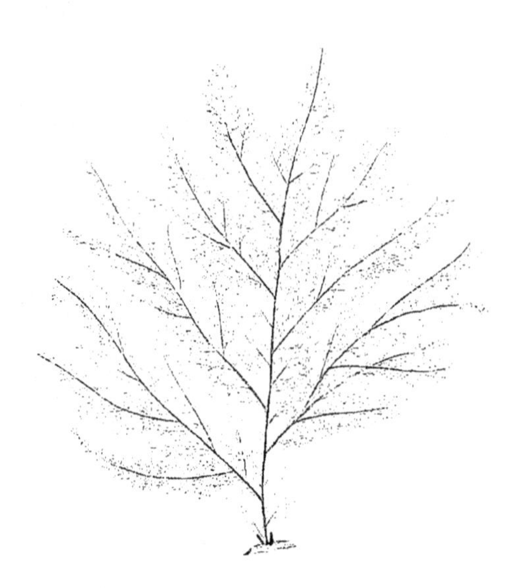

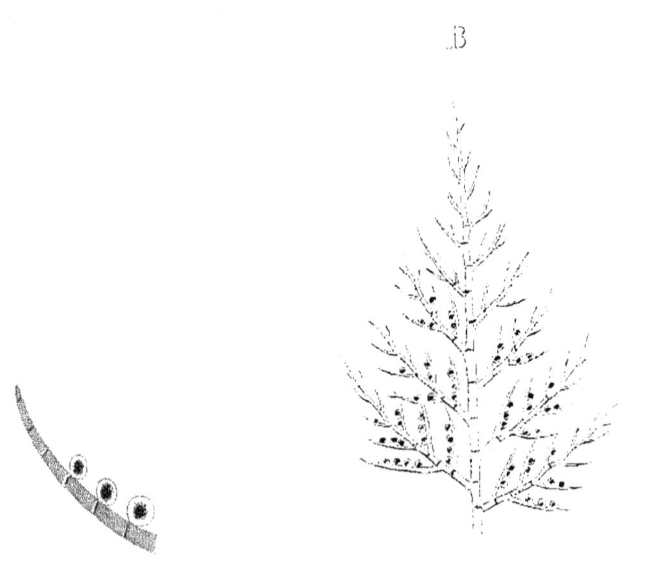

# CONFERVA ROSEA.

C. filamentis decompoſito pinnatis tenuiſſimis; ramis ramuliſque alternis, approximatis; diſſepimentis contractis; articulis oblongis, capſulis ſecundis ſub-globoſis.

Conferva roſea. Eng. Bot. t. 966.

Ceramium roſeam. Roth. Cat. Bot. II. p. 182.

On Planks and Fucus veſiculoſus in the River Yare, about Yarmouth Bridge, and on Rocks in the Sea, near Swanſea.

---

MY friend D. Turner has juſtly remarked to me that "it may be conſidered a ſtriking inſtance, how little the genus Conferva has been attended to by botaniſts, that above twenty years age, Mr. Wigg gathered the preſent ſpecies at Yarmouth, and preſerved ſpecimens of it in his Herbarium, which was ſo often viſited; but that till Mr. Sowerby found it there in 1797, and I, on ſending a plant of it to Dr. Roth, was informed of its being his Ceramium roſeum, no author of this country ever noticed it." That ſuch has been the caſe with many other ſpecies, I have already had occaſion to mention in this work, and is by no means a matter of aſtoniſhment, but the preſent conſidered as to its beauty, can hardly fail of attracting the moſt indifferent obſerver, and regarded as to its habit and mode of growth is ſo different from all the reſt, that no botaniſt could ever confound it with any common ſpecies. The root of C. roſea is a ſmall expanded diſk, which gives riſe to ſeveral ſtems, from one and a half to three inches in length, pinnated from their baſe with numerous alternate branches, which are again repeatedly ſubdivided in the ſame manner, ſo that as they approach the ſummits, they have a very cluſtered appearance; in their thickeſt parts they are nearly as fine as the hair of the human head, and ſo extremely fine towards their apices, as to be ſcarcely viſible. From the great tenuity of the ſhoots the ſubſtance of the whole is peculiarly flaccid, on which account it is difficult to expand it properly, but the ramuli, when floating in water, reſemble beautiful feathers. The joints are nearly oblong, and filled with a red fluid, which, after the plant has been immerſed ſometime in freſh water is given out, and ſtains the paper in drying. When perfectly freſh, the color of the whole is a roſy hue mixed with brown, uniform throughout, except that the leading branches are darker than the reſt.— It is not till after it has been expoſed to the air, or kept in freſh water, that the joints become pellucid, as deſcribed by Dr. Roth. The capſules are in general very numerous, and arranged on the upper ſides of the ramuli, nearly globoſe, very minute, and of the ſame color as the frond: when dried, hardly any ſpecies adheres more firmly to paper or glaſs.

A. C. roſea, natural ſize.
B. branch of ditto, magnified 2.
C. ſmall piece of ditto 1.

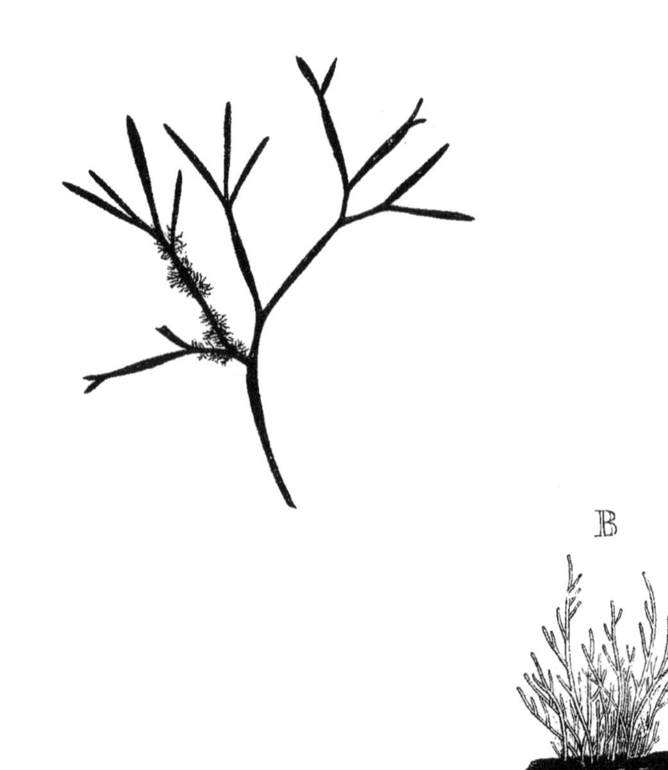
C
B

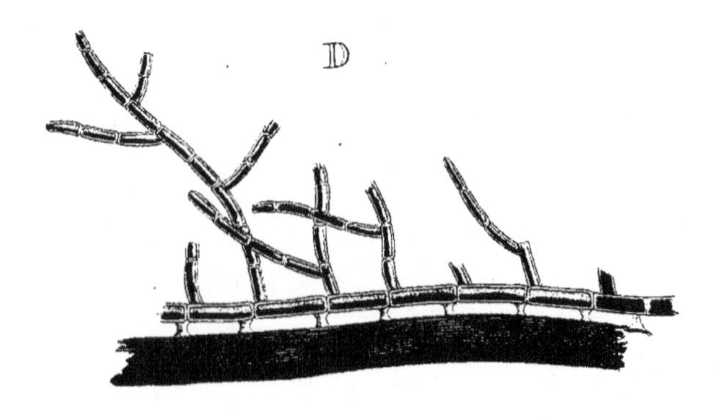
D

# CONFERVA REPENS.

C. filamentis minutis repentibus densè implexis; ramis ramulifque fubfecundis, diffepimentis parum contractis; articulis cylindricis.

C. marina per brevis villofa & cirrofa. Dill. Mufc. p. 23. t. 4. f. 21 ?

In the Sea paralitical on Fuci—At Yarmouth and Dover.

---

THIS delicate little parafite is found not unfrequently in the autumnal months attached to fucus lumbricalis, radiatus, & crifpus, & Conferva elongata. It grows in fmall clofely matted patches, and invefts the plants on which it grows in a very peculiar manner, as was firft pointed out to me by my friend Jofeph Woods, jun. to whom I am indebted for the drawings A & D, the latter of which is made upon a larger fcale than it really appeared under the microfcope, in order to fhew more clearly, than is otherwife poffible, its truly repent filament. He remarked that the branches rife only from thofe joints of the main ftem in which there is a radicle, but that they are never oppofite, and generally at the oppofite extremities of the branch. The length of the filaments feldom exceeds three or four lines; in thicknefs they are equal to the hair of the human head. Their whole length is befet with minute branches flightly incurved, almoft patent, accuminated at their apices, pointing upwards, and difpofed on the fame fide of the ftem, as are moft frequently the ramuli alfo. The joints are cylindrical, rather long, and flightly contracted at the diffepiments, which are pellucid. The color of the whole is a ferruginous red, inclining after it is dried to tawny. It adheres, though not firmly, to both glafs and paper. I have put a mark of doubt to the reference above quoted to Dillenius, becaufe I have not yet feen his collection, and he fays in his defcription, that the color of the plant is olivaceous or dark green. This circumftance, however, does not prevent my believing that he really meant this fpecies, and I would, on that account, have named it C. cirrofa, had not the name been previoufly given to a different one by Dr. Roth, in the fecond volume of his Catalecta. May not the prefent be Hudfon's C. fulva, the defcription of which, in the Flora Anglica, is unfortunately fo fhort, that unlefs any authentic fpecimen of it exifts, which I believe there does not, it will always be impoffible to tell what he meant by that name.

A. C. repens, natural fize, growing on Fucus lumbricalis.

B. Ditto magnified 3.

C. Ditto 1.

D. Ditto, on a larger fcale than it appeared in the microfcope.

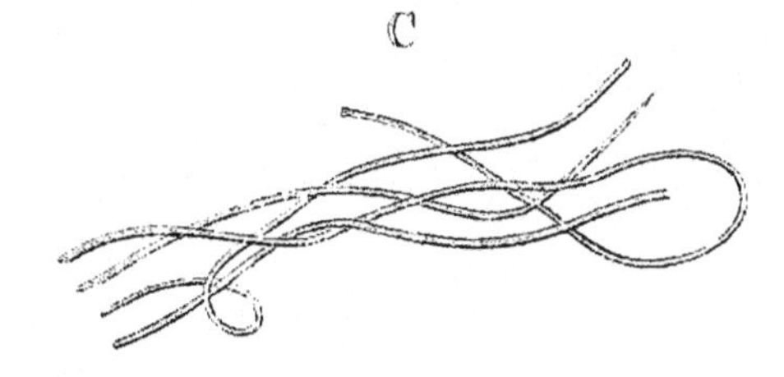


D

*Conferva myochrous*

Publish'd by L.W. Dillwyn Nov 1 1802.

# CONFERVA MYOCHROUS.

C. filamentis denſiſſimè implexis ramoſis; ramulis ſimplicibus ſubſecundis binis incurvis.

In Alpine Torrents at Beddgelert, and the lower regions of Snowdon. Dawſon Turner, Eſq.

---

FOR the following account of this ſpecies, which I believe to be entirely diſtinct from every other heretofore deſcribed, I am indebted to my friend Dawſon Turner, through whoſe indefatigable exertions it has been diſcovered, and to whom I am alſo indebted for the drawings A and B.

" This Conferva, in the month of July laſt, was extremely abundant in many of the torrents that flowed from the immenſe mountains which ſurround the beautiful vale of Beddgelert in Caernarvonſhire: it grew upon their rocky beds, matting the ſtones often, to a conſiderable extent, with a velvety covering, three or four lines in diameter, which, when taken out of the water, might aptly be compared to the ſkin of a mouſe. Its color was a dark gloſſy brown; its ſubſtance ſoft to the touch; its filaments ſo cloſely matted together, as to form almoſt an inſeparable maſs. Thoſe which I was able to detach, were ſeldom more than half an inch long, but I never was fortunate enough to find any with a root. Examined under a microſcope, their color appeared a pale ſubdiaphanous reddiſh brown, and there were in ſome ſpecimens, faint appearances of ſepta, but they were no where ſo evident as to warrant the inſerting them in the figure. The mode of ramification in this plant is very ſingular; ſome of the filaments being apparently quite ſimple, as in figures C & D. others twice or thrice trunked with patent dichotomies, and, as in figure B. beſet with pairs of ſimple incurved acuminated ramuli, arranged almoſt entirely on one ſide. Theſe latter are ſufficient at once to diſtinguiſh, and indeed the ſize of its filaments, which are as fine as the fineſt wool, will always keep it ſeparate from C. amphibia, the only ſpecies

I know to which it bears any ſtrong reſemblance. Some parts of Dillenius' deſcription of his 'Conferva mucoſa confragoſis rivulis inaſcens' ſo exactly correſpond with the preſent plant, that it may juſtly be doubted, whether, when he wrote his account of that ſpecies, he did not blend two or three together: as, however, Mr. Hudſon has referred that ſynonym to a different plant, and all ſucceeding botaniſts have followed him in ſo doing, it is not worth diſcuſſing the queſtion."

A. C. myochrous, natural ſize.
B. C. Ditto magnified 2.
D. Ditto 1.

B

*Conferva limosa*

# CONFERVA LIMOSA.

C. filamentis ſimplicibus, tenuiſſimis, brevibus, mucoſis, denſiſſimè compactis caeruleo-virideſcentibus lubricis; diſſepimentis indiſtinctis.

C. gelatinoſa, omnium tenerrima & minima aquarum limo innaſcens. Dill. Muſc. p. 15. T. 2. f. 5. Ray, Syn. p. 477.

On the muddy edges of Rivers, Ditches and Ponds.

---

THOUGH hardly any Conferva is more abundant than the preſent, eſpecially in Spring and Autumn, it appears neverthleſs to have remained unarranged in the ſyſtem, ſince the days of Dillenius and Ray.—It generally grows upon the mud, left at the edges of pools or ditches, preſenting to the naked eye, except immediately at the margin where it is fibrous, a widely expanded, thin, ſhapeleſs gelatinous maſs, reſembling a tremella, of a verv dark and gloſſy hue; ſometimes too it floats upon the ſurface of the water, and is conſpicuous by its dark green velvety appearance. In either caſe, the only mode to examine it is to carry it home, without allowing it to dry, and put it in a pan of water, where, though when firſt immerſed, its filaments are ſo thickly matted that they cannot be diſentangled, yet in the ſpace of a night it will ſhoot out an immenſe quantity of threads, viſible to the naked eye only from their number. The aid of a microſcope is neceſſary to obſerve them properly, and, thus examined, they preſent a curious appearance, for their length is not more than half an inch; they are obtuſe at each end, and lie croſſing each other without any apparent order—ſome indeed ſcem even to be wholly unconnected with the reſt. If the higheſt power of a good glaſs be applied, they ſcem to be jointed in an irregular manner, but this apparent irregularity is probably occaſioned by the want of ſufficient magnifiers, which, if we poſſeſſed, I am of opinion we ſhould find that the length of the joints is about equal to their breadth,as I have often faintly diſcerned two or more contiguous joints of theſc dimenſions. When the interior ſubſtance has collapſed by drying, if carefully examined, their tubular ſtructure may be obſerved. Dillenius's deſcription is ſo good, that I think it is impoſſible to miſtake him; he has publiſhed no repreſentation of it, aſſigning as a reaſon that ſince its parts elude the ſight, it would be raſhneſs to attempt a figure. There is however a rough pencilled ſketch among the original drawings in the extenſive library, ſo happily for ſcience, belonging to Sir Joſeph Banks, which merely repreſents a number of fibres lying together without any order.

The growth of this plant is aſtoniſhingly rapid, ſo that I have obſerved a very ſenſible difference in the length of its filaments in half an hour, and to this, and their extreme minuteneſs, which allows the ſlighteſt motion of the water, in which they are examined, to affect them, I attribute the motion obſerved by M. Adanſon, and deſcribed in an excellent paper in the Hiſtoire de l'Academie Royale des Sciences for 1767, page 75. Here M. Adanſon relates the following diſcovery, which, though I have not been ſo fortunate as to ſee the actual diviſion of the filaments he mentions, ſeems, from many appearances I have obſerved, extremely probable, and highly deſerving of further attention: "Lorſque ces filets ſont parvenus à leur dernier terme d'accroiſement, qui excède rarement trois lignes, alors le dernier nœud, qui n'a guère qu'une demi-ligne de long, s'en ſepare et s'alonge; ſes deux bouts s'arrondiſent, et il devient abſolument ſemblable à celui dont il s'etoit ſeparè, et capable d'en produire à ſon tour de nouveaux."

Dr. Roth's Conferva velutina ſeems from his account* of it to reſemble the preſent ſpecies both in ſubſtance and mode of growth, but ſpecimens from himſelf, and his deſcribing it 'filamentis ramoſis' prove them to be different.

A. C. limoſa, natural ſize.

B. Ditto magnified 1.

* Cat. Bot. I. p. 166.

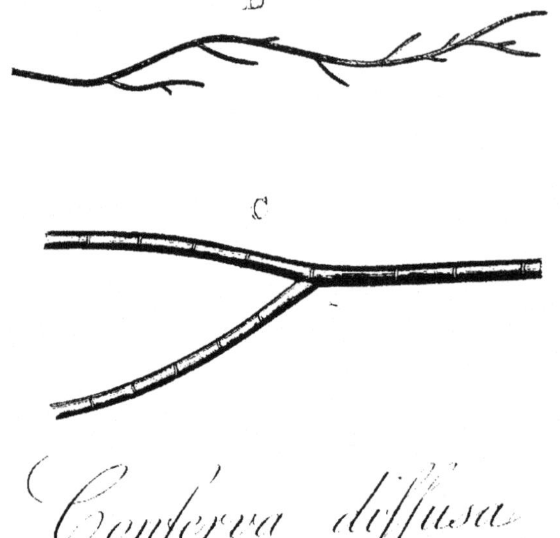


*Conferva diffusa*

Published by L. W. Dillwyn June 1. 1803.

# CONFERVA DIFFUSA.

C. filamentis ramofis diffufis; ramis fub-dichotomis flexuofis remotis; ramulis brevibus approximatis apice obtufis; diffepimentis pellucidis; articulis longiufculis.

Conferva diffufa. Roth, Cat. Bot. II. p. 207. t. 7.

On Rocks in the Sea near Swanfea.

---

Dr. ROTH, who firft defcribed this fpecies, informs us that it was difcovered growing on decaying wood and rocks at the Helder, by his indefatigable friend, Profeffor Mertens; to whofe pencil we are indebted for the figure of it in the fecond fafciculus of the Catalecta Botanica. It grows in loofely-entangled bundles, varying from two to fix inches in length, of a pale-green color, and more rigid nature than moft of its congeners; fo that, when drawn out of the water, its filaments do not collapfe. The root is a minute callus; each filament is in fize nearly equal to horfe hair; forked near its bafe, and afterwards repeatedly dichotomous, at remote, but irregular intervals, with alternate, flexuofe, rigid branches, often entangled almoft as much as thofe of Fucus plicatus. The ramuli are numerous, fhort, folitary, and fimple: fometimes placed alternately, but more frequently two, three, or four on the fame fide, and uniformly blunt at their apices: they originate at the diffepiments, which are pellucid. The joints are long, and cylindrical while frefh; but, in drying, generally contract in a very curious manner, as is reprefented in Cat. Bot. t. 7. C. & D.—In which alfo at B a number of fmall appendages are introduced. There were many agreeing with them in every thing but color on the plants I found at Swanfea, which proved on examination to be feedlings of Conferva rubra; a fpecies which, as well as many Polypi, often infefts this plant. I can hardly take a better opportunity of obſerving, what I truſt I may be allowed to obſerve without fear of being con-ſidered guilty of detraction, that almoſt all the plates of Conferva in the Catalecta Botanica are copied from plants, either in a dry ſtate, or which have been dried. They are not therefore in general applicable to the ſpecies examined while recent. Many of the deſcriptions labor under a ſimilar diſadvantage, from the learned author's reſiding at ſo great a diſtance from the ſea. It was neceſ-ſary to mention this circumſtance on my own account; becauſe, had it not been noticed, it muſt have been thought that the figures in this work contradict thoſe of Dr. Roth, and, ſtill more, becauſe nothing would be more likely to miſlead a young botaniſt. The fructification of C. diffuſa has not yet been diſcovered; it adheres, when dried, very ſlightly to paper, and not at all to glaſs.

A. C. diffuſa, natural ſize.

B. Ditto, magnified 4.

C. Ditto, 2.

A

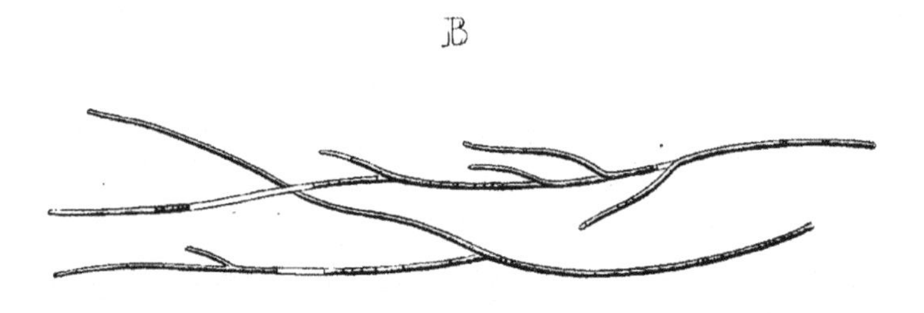


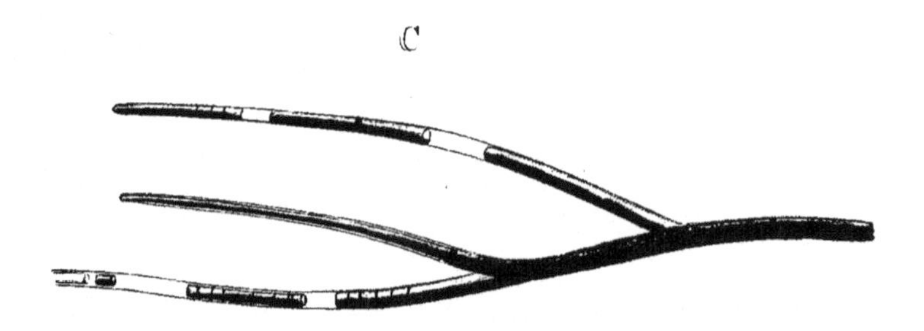

# CONFERVA DISTORTA.

C. filamentis ramofis articulatis fafciatis; ramis ramulifque diftortis; diffepimentis obfoletis, articulis brevibus.

C. diftorta. Fl. Dan. t. 820.

In a boggy Pool on Sketty Burroughs near Swanfea.

---

ALTHOUGH, in the prefent ftate of our knowledge of Convervæ, it is impoffible with certainty to fay which fpecies are moft rare, or which moft common, I cannot but think that the prefent has a claim to be confidered as one of the fcarceft of the tribe. Muller, whofe figure in the Flora Danica is excellent, was the firft who noticed it; and he appears to have found it only in one place, and there but very fparingly.

I have a German fpecimen, through the kindnefs of my friend Dawfon Turner, from Dr. Roth, under the apt name of Ceramium natans; and I do not know that it has been found by any other botanift, till I fortunately met with it in fmall quantity laft autumn in fome boggy pools on Sketty Burroughs near Swanfea. It grows parafitically in fhort thick tufts on decaying grafs; attached to fmall pieces of which it frequently floats on the furface of the water. The root I have not been able to difcover; its filaments are generally about half an inch in length, extremely flender, and of a beautiful dark green color, varying to a lighter hue as they approach to decay. The branches are feldom numerous, but have a very peculiar twift at their ramification, from which is derived the fpecific name of the plant, and which is its greateft peculiarity. Muller, though his figure abounds with tranfverfe lines, defcribes this fpecies 'filamentis inarticulatis' and hence appears to have difcovered a difference between thefe lines and true diſſepiments; but although 'faſciæ,' ſimilar to thoſe mentioned under C. muralis, with which this ſpecies has a ſtrong affinity, frequently appear, diſſe-piments may alſo be diſcovered; and indeed I conſider the remarkably abrupt manner in which the juices are frequently ſeen to have collapſed in ſome others as well as the preſent ſpecies, as a clear indication of their exiſtence. To the naked eye the ſize of the filaments, their mode of growth, and color; and, under the microſcope, their ſingular ramification, at once diſtinguiſh C. diſtorta from all other ſpecies. It adheres, when dried, to either glaſs or paper.

A. C. diſtorta, natural ſize.

B. Ditto, magnified 3.

C. Ditto, 1.

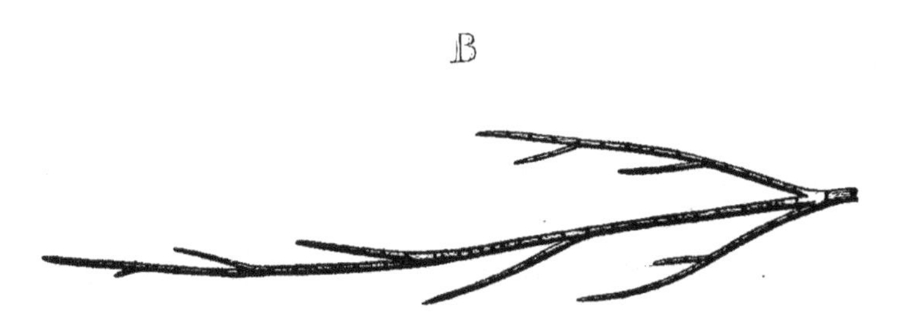
B

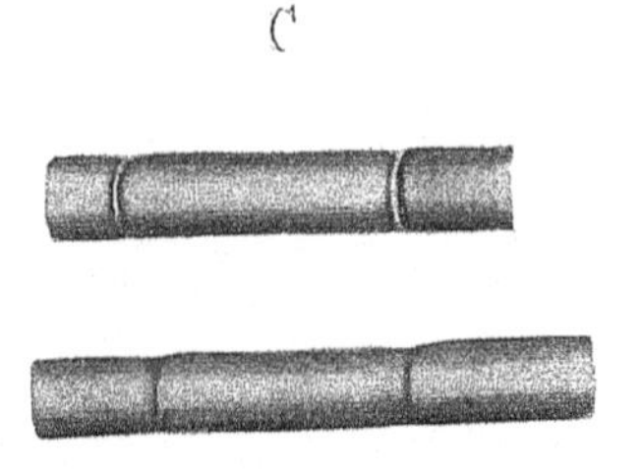
C

# CONFERVA RUPESTRIS.

C. filamentis ramofiffimis fafciculatis ftrictis virgatis adpreffis, apice truncatis: diffepimentis parùm contractis, cryftallinis; articulis longis, cylindricis.

C. rupeftris. Sp. Pl. p. 1637. Fl. Ang. p. 601. Fl. Scot. p. 994. With. IV. p. 140. Fl. Dan. t. 948. Roth, Cat. Bot. II. p. 208. Fl. Germ. III. pars 1. p. 516.

C. glauca. Roth, Cat. Bot. II. p. 208. t. 6.

C. marina trichodes ramofior. Dill. Hift. Mufc. p. 28. t. 5. f. 29.

C. marina trichodes f. mufcus marinus virens tenuifolius. Ray. Syn. p. 60.

On Rocks and Stones in the Sea, common.

---

THIS elegant fpecies, one of the moft common ornaments of our fhores, appears to have been longer and better known than moft of the Confervæ. Its root, a fmall callus, gives rife to a number of dark green filaments, fomewhat rigid to the touch, which are fo repeatedly branched, that each of them affumes a bufhy appearance; the mode of ramification is irregular, fome of the branches being alternate, and fome oppofite; while, towards the fummit, three or four are frequently difpofed without interruption on the fame fide.—All of them are erect, and remarkable for their ftraitnefs, as well as for being placed very clofe to each other: the ends are always blunt, and generally fo much fo as to have a truncated appearance; but in fome fpecimens, this is more ftrikingly the cafe than in others. On this, and other fmall variations, to which this plant is liable, Dr. Roth, who does not appear ever to have had an opportunity of examining it when frefh, has founded another fpecies under the name of C. glauca, which, upon the authority of ſpecimens ſent from Profeſſor Mertens to D. Turner, I have felt no heſitation in uniting with the preſent, nor can I ſee ſufficient grounds to deſcribe them even as ſeparate varieties.

The diſſepiments are a little contracted, and generally quite colourleſs; but, before the plant is expoſed to the air, or the juices at all collapſed, they on the contrary appear darker than the other parts of the filaments, the joints are cylindrical, and of a deeper color towards their extremities; their length, though ſubject to ſome variation, even in the ſame branch, is ſeldom leſs than double their width.

Dillenius's "Conferva fluviatilis trichodes, extremitatibus ramoſis," which moſt authors have followed Hudſon in making a variety of this plant, appears to be a diſtinct ſpecies:—it is the Ceramium aſperum of Dr. Roth. In drying C. rupeſtris retains its beautiful green color; but the joints contract alternately in a curious manner, as figured in the Catalecta Botanica:—it adheres to neither glaſs nor paper.

A. C. rupeſtris, natural ſize.

B. Ditto, magnified 3.

C. Ditto, 1.

B

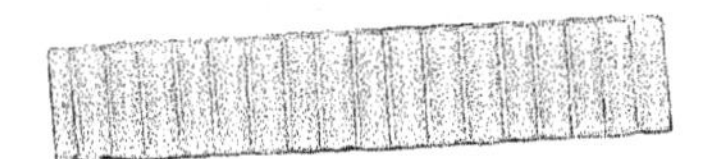

*Conferva pectinalis*

Published by L.W. Dillwyn June 1 1803.

F. Sansom sculp.

# CONFERVA PECTINALIS.

C. filamentis, ſimplicibus, pellucidis, fractis, acuminatis; diſſepimentis ſæpè ſolutis; articulis breviſſimis medio cryſtallino-pellucidis.

C. pectinalis. Muller in Nov. Act. Pet. III.

C. bronchialis. Roth. Cat. Bot. I. p. 186. Fl. Germ. III. p. 520.

In rivers and ſtagnant waters, adhering to decaying wood and vegetables.

---

MULLER, who firſt found this ſingular ſpecies, and publiſhed an excellent figure of it in the paper above referred to, obſerves that it is abundant in the ditches about Pyrmont. Dr. Roth alſo remarks that it is not rare in thoſe near Vegeſack; and though not one of our moſt common ſpecies, it frequently occurs in ſimilar ſituations in many parts of this country, eſpecially in the neighbourhood of London, where it is very plentiful, and where, early in the ſpring of laſt year, I firſt found it in the company of my friend, Joſeph Woods, junr.

The filaments are of a dirty green colour; ſeldom exceeding half an inch in length, and to the unaſſiſted eye, reſemble decayed vegetable matter. When entire they gradually taper to a point, and, as Muller obſerves, bear ſome reſemblance to the antennæ of a lobſter, but I could never obſerve the appearance of cylindricity repreſented in the figure of it given by that botaniſt. The diſſepiments are very conſpicuous, and at theſe the filaments frequently break; the parts remaining connected at only one extremity; which, when it repeatedly takes place, gives the plants ſo much the appearance of *flocculoſa* as to make it ſomewhat doubtful whether the ſpecies are diſtinct: the joints are very ſhort, and appear coloured towards each end by a green fluid, which, ſoon after the plant is taken from the water, and as it approaches to decay, collapſes, ſometimes forming into ſmall globular maſſes, and ſometimes diſappearing entirely.

C. pectinalis may be readily distinguished from its Congeners by the remarkable change it undergoes when dried; it then turns to a greenish ash-colour, and shines as if covered with gum-water. In that state it adheres firmly to either glass or paper.

A. C. pectinalis, magnified, 2.

B. Ditto, 1.

A

B

C

(

# CONFERVA ATRO-VIRENS.

C. filamentis rigidiufculis ramofis; ramis divaricatis, fub-fecundis, utrinque attenuatis, apicibus obtufiufculis; diffepimentis pellucidis; articulis breviffimis tripunctatis.

On the wet Rocks, forming the banks to the Dylais River, near Neath.

---

THIS fingular, and hitherto unobferved fpecies, abounds on the dripping rocks which conftitute the banks of the River Dylais, near where it forms the romantic cafcade, fo well known to thofe who have vifited the highly picturefque neighbourhood of Neath, in Glamorganfhire. It grows in thick bufhy tufts, of a blackifh green color; and from its rigid nature, is liable at firft fight to be miftaken for one of the Mufci, with which it is not unfrequently mixed. The root appears to be a very minute callus. The filaments, from a quarter to half an inch in length, are divided into numerous branches, which are difpofed without any apparent order, though feveral together are moftly on the fame fide of the main ftem, with which they form a very obtufe angle: they taper in fome degree both towards their origin and apex, but terminate rather bluntly. The diffepiments are pellucid: the joints very fhort, not much exceeding the diffepiments in length, and as it appears under the microfcope, compofed of three granules, which, not having been able to find any other, I conclude are the fructification of the plant.

There is no danger of its being confounded with any other fpecies.

When dried its color becomes rather darker, and in that ftate it will not adhere to glafs or paper.

A. C. atro-virens, natural fize.
B. Ditto magnified 3.
C. Ditto 1.

A

B

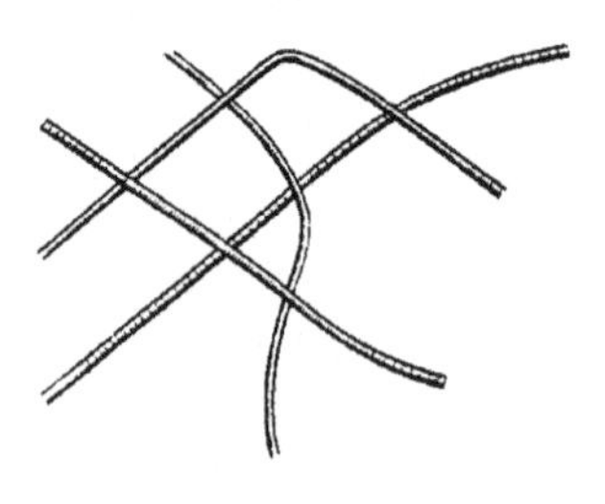

*Conferva decorticans*

# CONFERVA DECORTICANS.

C. filamentis ſimplicibus, tenuiſſimus, denſiſſimè contextis, cœruleo-viridefcentibus; diſſepimentis obſcuris; articulis brevibus.

On damp walls and ſtones not uncommon.

---

THIS ſpecies, which appears hitherto to have eſcaped obſervation, is by no means unfrequent on walls and ſtones much expoſed to moiſture. I firſt detected it mixed with C. muralis, on the pump facing Stationer's Hall, in London, and ſince in ſimilar ſituations in ſeveral of the Weſtern counties. It grows in large glaucous patches, ſo intimately woven as to peel off in flakes, bearing a conſiderable reſemblance to a piece of ſilk or ribbon: its filaments, which it is impoſſible to diſentangle ſo as to aſcertain their length, are extremely ſlender, of a deep glaucous color, and ſome of them are very ſlender: diſſepiments may be obſerved regularly diſpoſed at diſtances about equal to the thickneſs of the filament.

From C. muralis it differs in its much greater tenacity, and darker color; from C. limoſa in the former, and its far different mode of growth; and from both theſe, and indeed all others, it may be diſtinguiſhed by its forming patches ſo denſely matted as to peel off in thin ſtrata, as is above deſcribed.

The ſurface is in general very ſmooth and gloſſy; but when the wall on which it grows is occaſionally waſhed by a ſtronger ſtream of water than uſual, as frequently happens at mills, its filaments are lengthened out, and the ſurface aſſumes a more ſhaggy appearance.

In drying, it does not appear to ſuffer any change, and adheres to both glaſs and paper.

A. C. decorticans, natural ſize.

B. Ditto, magnified, 1.

*Plate 27*

A

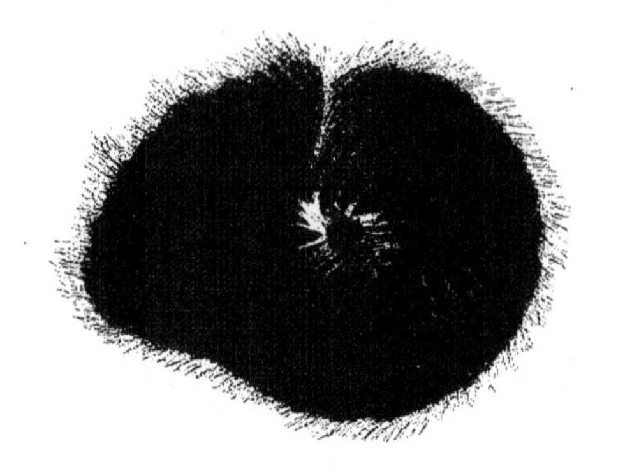

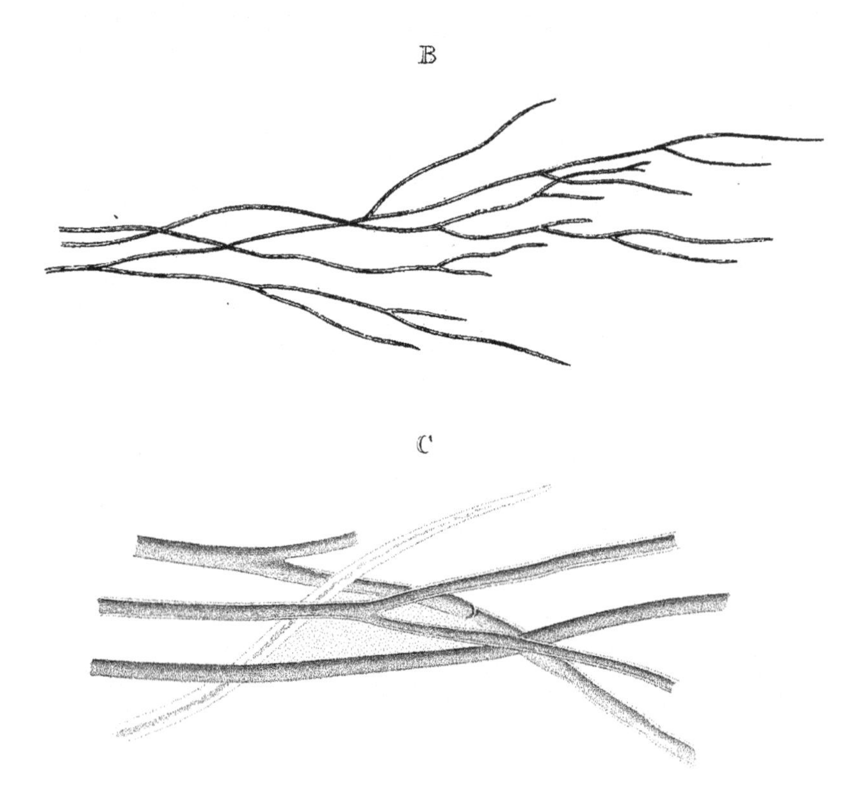

# CONFERVA COMOIDES.

C. filamentis tenuibus, ramofis: ramis fparfis, remotiufculis, apice acuminatis: diffepimentis parum contractis, ferè obfoletis.

On feveral of the marine algæ and rocks in the fea at Swanfea.

---

THIS fpecies I believe to be extremely common on our fhores, though it appears hitherto to have been entirely overlooked, or perhaps confidered as the feedling of C. littoralis, to which it bears fo great a refemblance that it is not without hefitation I have ventured upon publifhing it as diftinct; though from repeated obfervation I have found its characteriftic marks fo conftant, that, if not fpecifically different, it muft at leaft be allowed to be a moft fingular variety: and, in the prefent ftate of our knowledge of thefe plants, I conceive nothing more can be expected from any author, nor indeed any thing be done more favourable to the advancement of fcience, than, by giving faithful figures and defcriptions of what we fee, to ftore up materials for future naturalifts to work upon. The naked eye may readily diftinguifh the two plants, by the fmaller fize of C. comoides, which feldom exceeds an inch in length, and its deeper color, either of, or approaching to, a purple brown. Under the microfcope their different ftructure is fuch, that I hope it will not be poffible to confound them. The prefent fpecies grows on marine ftones and algæ, and frequently fo covers the round pebbles which abound among the rocks with its flender hair-like tufts, lying one over the other, as to give them a ftriking refemblance to the head of an infant. The branches are rather irregular, and not fo numerous as in C. littoralis; but, as in that fpecies, they originate at very acute angles, and are acuminate at their apices. The diffepiments being extremely faint, it is almoft impoffible to afcertain the fize of their joints, but their length always appears much to exceed their width; as, where contractions occur, which is generally the only mark by which the dissepiments can be discovered, the filaments gradually and slightly diminish for a considerable distance towards them. In drying, this plant changes to a greenish grey color, and adheres both to glass and paper.

A. C. comoides, natural size, growing on a pebble.
B. Ditto magnified 3.
C. Ditto Ditto 1.

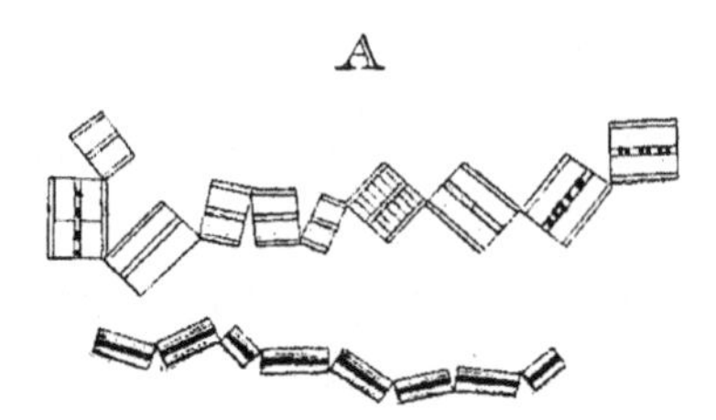
A

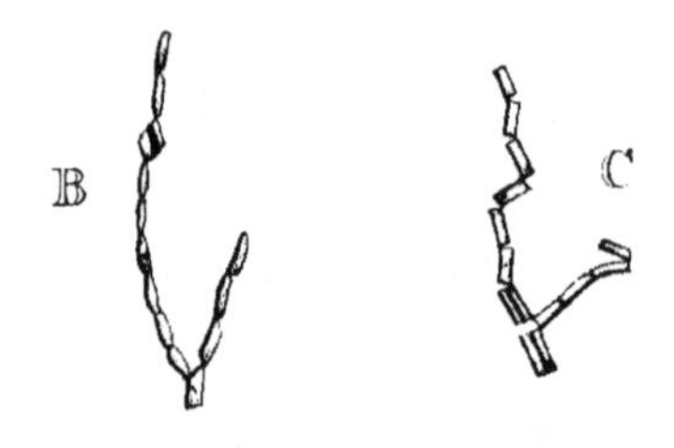
B
C

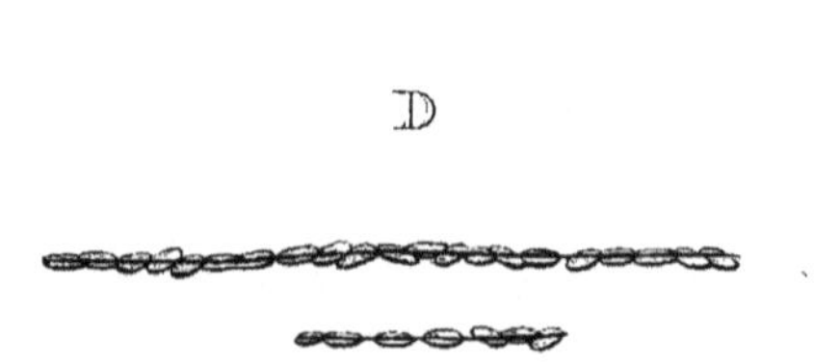
D

# CONFERVA FLOCCULOSA.

C. filamentis ſub-ſimplicibus compreſſis, minutis; diſſepimentis ſolutis; articulis priſmaticis, alternatim refractis.

C. flocculoſa. Roth. Cat. Bot. I. p. 192. t. 4. f. 4. & t. 5. f. 6. Fl. Germ. III. pars 1. p. 523.

In Pools, Ditches, and Slow Streams, adhering to other Confervæ, and to decaying vegetables.

---

THIS ſingular plant was found for the firſt time in Britain by my friend Joſeph Woods, junr. and myſelf, growing on decaying vegetables in a pool on Hampſtead Heath, ſince which time I have obſerved it in various other places. Its ſtructure is ſo extraordinary, that notwithſtanding the figures and deſcriptions in the Catalecta Botanica, and my own repeated obſervations, I can hardly now allow myſelf to aſſign it a place among the perfect productions of nature. I think it beſt however to ſubmit a figure of it to the Botanical world, and ſhall be happy to abide by their deciſion. At firſt I conſidered it as C. pectinalis broken to pieces, but a little obſervation rendered that idea inadmiſſable. It certainly has very much the appearance of a broken plant; but J. Woods, junr. has obſerved it in a ſtate figured at C. in which the joints cannot be ſo diſpoſed as to make the two parts of the line, which one might otherwiſe imagine continued originally the whole length of the plant, coincide.

It is a very ſmall ſpecies, ſeldom exceeding one-fourth of an inch in length, and varying in color from a pale to a greeniſh brown. The filaments are rarely branched; their form is not eaſily aſcertained, but they have always appeared to me to be very much compreſſed; and the joints, only adhering to one another by ſingle points, look like a ſtring of parallelograms united at the corners. Each joint has a double line running through the middle of it, and ſome very faint tranſverſal bands frequently appear; in ſome caſes however, as at B, &c. this line is either entirely wanting, or has eſcaped the power of my glaſs.

C. flocculoſa is ſubject to ſome variations, of which all that have hitherto been obſerved are noticed in the plate. A. repreſents the plant as it generally appears, and indeed though I frequently examined, I ſaw it in no other ſtate for ſome months, but on the 23d of May, 1802, my above mentioned friend found it in the New River, as repreſented at B, and he afterwards obligingly communicated a drawing of ſome that he found varying ſtill more from its general appearance, which is given at D. Its favourite ſituation is on C. glomerata, which about London is ſeldom to be met with without it.

It adheres well to either glaſs or paper.

A. C. flocculoſa, magnified 1.

B. C. D, Ditto, ditto 2.

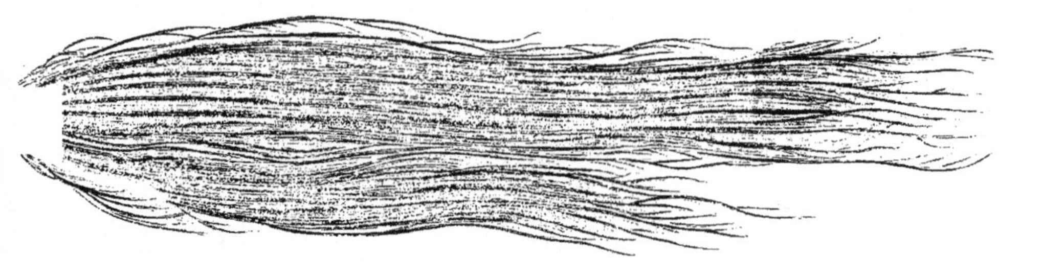

A

B

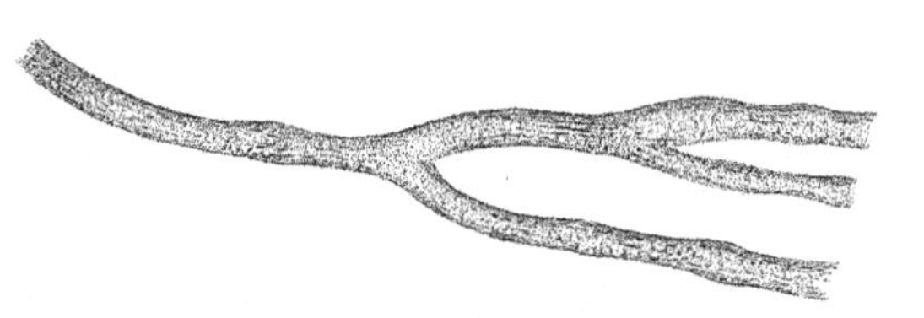

C

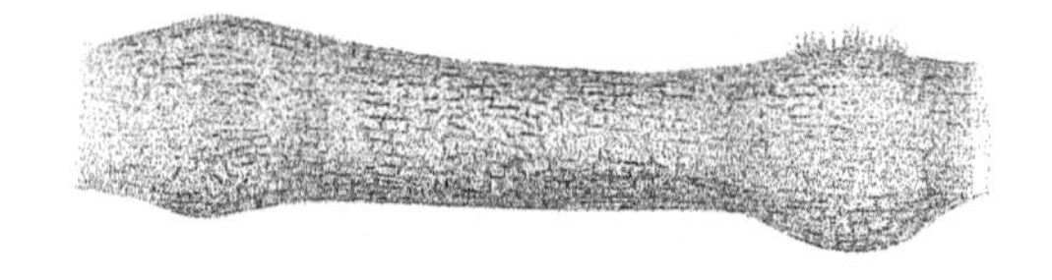

# CONFERVA FLUVIATILIS.

C. filamentis ramosis rigidiusculis; ramis ramulisque subalternis utrinque attenuatis; dissepimentis torosis, verrucosis; articulis longis bifariam dilatatis.

C. fluviatilis. Sp. Plant. p. 1635. Fl. Ang. p. 597. Scot. p. 985. With IV. p. 134. Roth. Cat. Bot. I. p. 201. Fl. Germ. III. pars 1. p. 528.

C. torulosa. Roth. Cat. Bot. I. p. 202. Fl. Germ. III. pars 1. p. 529.

C. fluviatilis nodosa, fucum æmulans. Dill. Musc. p. 37. t. 7. f. 4. 8.

C. fluviatilis lubrica, setosa, equiseti facie. Dill. Musc. p. 39. t. 7. f. 47.

In rapid and rocky streams, in Yorkshire, Cumberland, and Westmoreland, Hudson. Common in the Western counties of England and in Wales.

---

C. FLUVIATILIS abounds in most of the rapid rivulets in Wales, and the West of England, growing in large masses, generally of a dull olive color, but sometimes varying to a greenish purple. The root is a small callus, common to several filaments, which are six or eight inches long, irregularly divided and sub-divided into branches and ramuli attenuated at both ends. The principal branches are about the thickness of common twine; but the ultimate ramuli are often as fine as the hair of the human head. Sometimes however its filaments are nearly simple, when they are shorter, thicker, and more rigid than those which are much branched. In this state it is most probably the C. fluviatilis nodosa, fucum æmulans of Dillenius, and the C. torulosa of Dr. Roth; but, as I have observed both appearances on filaments growing from one root, I cannot consider them as distinct species. The dissepiments are swollen, so as to appear very evident to the naked eye, and are generally beset with two, three, or sometimes four hairy tubercles, which are perhaps in some manner connected with the fructification, though no seeds have hitherto been detected in them.

The joints are oblong, narroweſt in the middle, and beautifully reticulated with dark colored veins: their length is about equal to eight times their thickneſs.

Mr. Turner and Mr. Sowerby found near Penzance, a gigantic variety of this plant, extending to two feet in length, and with the joints of its branches quite obſolete.

In drying, the colour becomes darker, and it will adhere ſlightly to paper, but not at all to glaſs.

The drawings of C. fluviatilis and gelatinoſa, were executed by Wm. W. Young, an ingenious artiſt at Swanſea.

A. C. fluviatilis, natural ſize.

B. Ditto, magnified 4.

C. Ditto, 2.

B

C

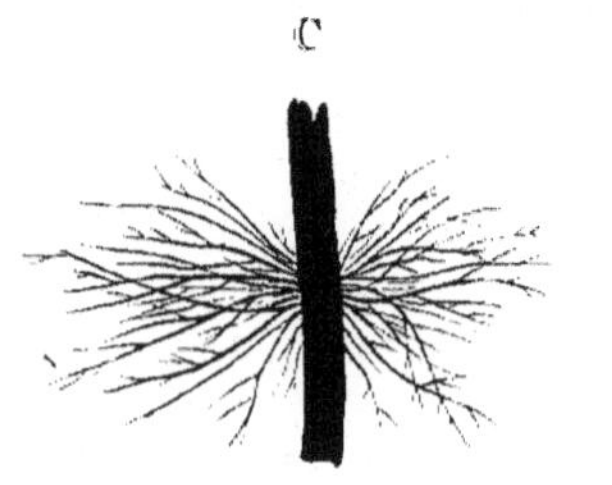

D

*Conferva nana.*

# CONFERVA NANA.

C. filamentis ramoſis minutiſſimis: ramis ramuliſque ſub-alternis acuminatis; diſſepimentis pellucidis; articulis cylindricis.

In the Wye, near Llanydloes, in Montgomeryſhire, and near Swanſea.

---

THE bottom of the rocky channel of the Wye, near Llanydloes, was on the 9th of laſt November covered with a ſoft down, which, on ſubſequent examination, proved to be the remains of ſome Conferva in decay; moſt probably the Ceramium cæſpitoſum of Roth, overgrown with the preſent extremely delicate paraſite. The minuteneſs of the filaments, which, in length, ſeldom much exceeded half a line, prevented me from aſcertaining their nature ſo fully as I could have wiſhed. Their color is pale brown, tinged with green, ſubdiaphanous under the microſcope. They appear to conſiſt of a ſimple ſtem, beſet at uncertain diſtances with alternate branches, which are again clothed with ſhort, ſimple, ſolitary ramuli, placed at ſmall diſtances from each other, moſt commonly alternate, though ſometimes two or more together are diſpoſed on the ſame ſide; all of them are finely acuminated: the diſſepiments are very apparent, and divide the filaments into joints, all of equal ſize, of which the length is about double the thickneſs. To the naked eye, this plant, like ſome of the moſt common ſpecies, appears, when taken from the water, like a mere maſs of decaying vegetable matter; its extreme minuteneſs might fairly induce a ſuſpicion whether it is in reality any thing more than the ſeedling of ſome known Conferva; and under this idea I ſhould have been unwilling to publiſh it as a new one, but that its ramification, and remarkably acuminated branches, render it quite unlike any other, with which I am acquainted, except C. littoralis, which is one of our largeſt ſpecies; and which, it may therefore fairly be preſumed, would not be ſo perfectly formed in ſo very minute a ſtate. It adheres to either glaſs or paper.

A. C. nana, natural ſize.

B. Ditto, magnified 3.

C. Ditto, 2.

D: Ditto,

Plate 37

# CONFERVA LITTORALIS.

Conferva filamentis ramofiffimis tenuiffimis flexuofis densè implexis; ramis ramulifque acuminatis: diffepimentis obfcuris; articulis cylindricis brevibus.

C. littoralis. Sp. Pl. p. 1634. Fl. Ang. p. 594. Fl. Scot. p. 979. With. IV. p. 130. Cat. Bot. I. p. 152.

Conferva marina capillacea longa ramofiffima mollis. Dill. Hift. Mufc. p. 23. t. 4. f. 19.

On Rocks and Fuci in the Sea, common.

---

THIS fpecies abounds on all the coafts I have yet examined, growing either on the rocks or larger fuci, particularly Fucus veficulofus. Its filaments, which are peculiarly thin, form oblong fafciculi, of a dull olive-green color, occafionally more or lefs tinged with yellow; and varying in length from fix to nine inches: they are much branched, and fo flender and flexible as to be affected by the flighteft motion of the water. Their fubftance is tender and foft, but ftill by no means inclining to gelatinous. Their mode of growth is fo entangled, that it is almoft impoffible to feparate them. In their native fituation they have a remarkably elegant appearance, being twifted together fo as to look like thick fhoots, the edges of which are, from the young branches, feathered in a moft beautiful manner. The branches are generally alternate, but fometimes oppofite, and iffue from the ftem at acute angles. They are always remarkable for their acuminated apices, upon which, and the fhortnefs of the joints, the ftrongeft characters of this plant depend. The diffepiments are nearly black, often appearing broken, and dividing the filaments into fhort joints; which, as in C. fracta and fome other fpecies, are frequently fwollen, and then affume a darker color. Some fpecimens

I gathered laſt ſpring in the river Yare at Yarmouth, from which the drawing at D was made, produced numerous globular capſules, ſcattered irregularly on the branches. In drying, the color of this plant becomes ſomewhat darker, and it adheres, though not very firmly, to both glaſs and paper.

A. C. littoralis, natural ſize.
B. Ditto, magnified 4.
C. Ditto ditto, 1.
D. Small piece of do. in fruit. 2.

B

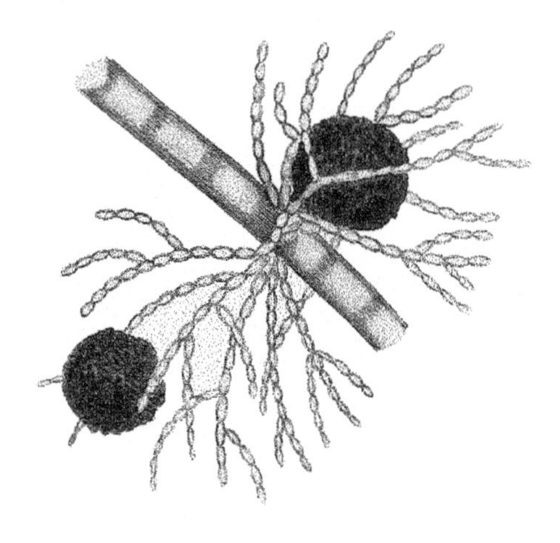

*Conferva gelatinosa*

Published by L.W. Dillwyn June 1.1803

F. Sansom sculp

# CONFERVA GELATINOSA.

C. filamentis ramofiffimis moniliformibus lubricis; ramulis tenuiffimus, penicilliformibus, fub-verticillatis, ramoffimis fructiferis; diffepimentis obfcuris; articulis breviufculis; capfulis fub rotundis polyfpermis

C. gelatinofa. Lin. Sp. Plant. p. 1635. Fl. Ang. p. 597. Fl. Scot. p. 986. With. IV. p. 134. Eng. Bot. t. 689.

Chara batrachofperma. Weis Gott. p. 33. t. 1.

Chara gelatinofa. Roth. Cat. Bot. I. p. 126.

Batrachofpermum moniliforme. Roth. Fl. Germ. III. pars 1. p. 480. Cat. Bot. II. p. 187.

Conferva fontana nodofa, fpermatis ranarum inftar lubrica major and fufca. Dill. Mufc. p. 36. t. 7. f. 42. Ray Syn. p. 62.

Common in clear ftreams.

---

IN the defcription of C. mutabilis it is remarked, that two out of the five plants which Linnæus united under this name, are certainly diftinct, and that the remaining three exhibit little other difference than that of fize of color. Obfervations fince made by D. Turner, Efq. and Jofeph Woods, junr. on the Herbaricum of Dillenius, prove that his No. 43, is marked only by its fmaller fize, and hardly deferves to be confidered a variety, and fome fpecimens gathered laft fummer by D. Turner, in Lyn Fynnon Velan, an Alpine Lake on Snowdon, leave it very doubtful whether 45 is not really a diftinct fpecies, its ftems being quite hard, and its monifiform appearance very faint; fo that I think it beft to defer mentioning it till it fhall fall more immediately under my own infpection in a recent ftate.

The root of this ſpecies is a black callus, the ſhoots are numerous, and when taken from the water collapſe together, ſo as to form a ſhapeleſs maſs, bearing a ſtriking reſemblance to frog's ſpawn, and ſo ſlippery, that the fingers can with difficulty hold it. It varies in length from one to ſix inches, and in color, from a dark purple to a blackiſh or yellowiſh green. When expanded in water the filaments are ſeen to be repeatedly branched, the branches diſpoſed without any regular order, and beaded in a very elegant manner. Under the microſcope theſe beads appear to be formed by ramuli either oppoſite or verticillate, repeatedly divided and ſubdivided into extremely ſhort and ſlender patent branchlets of nearly equal lengths; among them, at, I believe, every ſeaſon of the year, fructification may be found conſiſting of minute globular blackiſh capſules, which, when highly magnified, may be obſerved to be compoſed of an immenſe number of grains. On this account principally Dr. Roth has made it a new genus under the name of Batracho-ſpermum, but, as already remarked in Engliſh Botany, the fruit of all the marine Confervæ is a polyſpermous capſule. Weis, as above quoted, has referred this plant to the Charæ, a tiribe of plants moſt ſtrikingly connected with the Confervæ.

In drying, it changes but little, and will adhere firmly to either glaſs or paper.

A. C. Gelatinoſa, natural ſize.

B. Ditto, magnified 4.

C. Ditto, ditto 1.

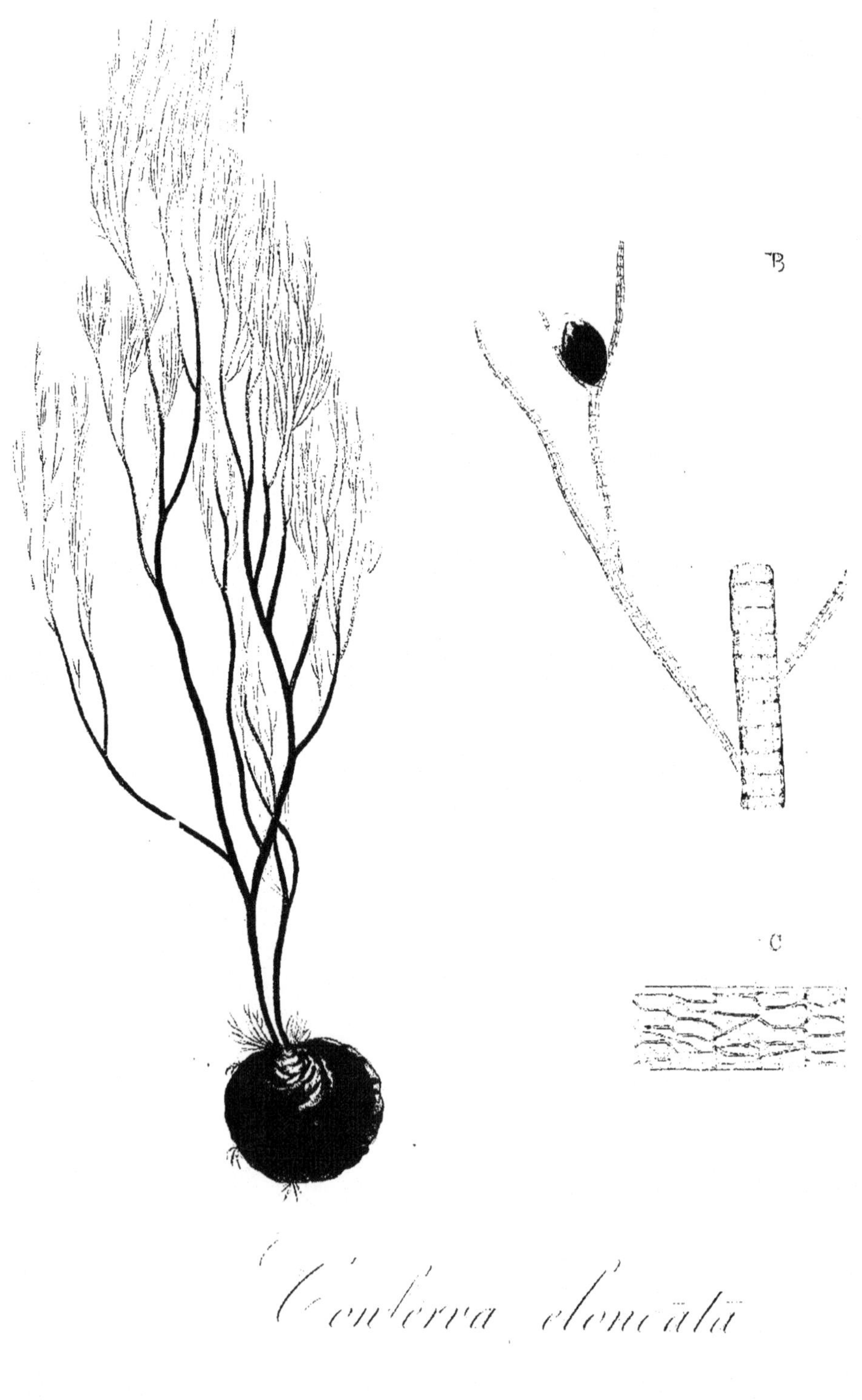


*Conferva elongata*

# CONFERVA ELONGATA.

C. filamentis ramofiffimis cartilagineis; ramis ramulifque elongatis, diffufis, fetaceis, venofis; diffepimentis obfcuris articulis breviffimis, capfulis ovatis feffilibus.

C. elongata. Fl. Ang. p. 599. With. IV. p. 137.

Fucus diffufus. Fl. Ang. p. 589. Lin. Tranf. III. p. 197.

On Rocks in the Sea, common.

---

C. ELONGATA, which in fize exceeds every other Britifh Confervæ, is extremely common on moft if not all our fhores, and I have frequently feen it adhering to oyfters in the London markets. It has been often, and not uncommonly, called the Lobfter-horn Conferva. Its root is an expanded callus; the frond in general folitary; the main ftem is as thick as common twine, and of a more cartilaginous and firm texture than in any other fpecies. The branches and ramuli are fetaceous, long, diffufe, and elegantly veined; under the higher powers of the microfcope the veins prefent a very remarkable appearance, being filled with a fluid, which, in drying, collapfes towards the middle, precifely as reprefented in the magnified filaments of Conferva glomerata (tab. 13.); and in them a few diffepiments may be here and there obferved, dividing them at uncertain and irregular diftances. The larger of thefe veins, or rather perhaps thofe which are difpofed on the furface, anaftomoze at the diffepiments, as if they were the origin of them, but a little obfervation fhews that they are quite independent of each other. The diffepiments are of a darker color than the reft of the filaments; the joints are very fhort, being feldom more in length than half their breadth. The capfules, found in the Months of July and Auguft, are fcattered rather fparingly on the ultimate branches; they are ovate, feffile, and in nature exactly refemble thofe of C.

coccinea; but befides them, C. elongata, in the early months of the fpring, as mentioned in the Synopfis of Britifh Fuci, p. 355. is fometimes covered with capfules fimilar to thofe of Fucus fubfufcus, with which plant and F. pinaftroides it has, in point of general habit, fo ftrong a refemblance, that they cannot be feparated without violence. Hudfon, by twice introducing this fpecies under different names in the Flora Anglica, has been the caufe of great confufion.

In drying, it affumes a darker color, and adheres very flightly to paper, and not at all to glafs.

A. C. elongata, natural fize.
B. Ditto magnified 3.
C. Small piece of Ditto 1.

B

# CONFERVA RUBRA.

C. filamentis ramofiffimis; ramulis fetaceis, apicibus furcatis; diffepimentis parum contractis; articulis in medio pellucidis; capfulis fubglobofis lateralibus.

C. rubra. Fl. Ang. p. 600. With. IV. p. 138. Eng. Bot. t. 1166.

C. nodulofa. Fl. Scot. p. 944.

Ceramium virgatum. Roth, Cat. Bot. I. p. 148. t. 8. f. 1. (excl. fyn. Hudf.) Fl. Germ. III. pars. 1. p. 461.

C. geniculata, ramofiffima lubrica longis fparfifve ramulis. Dill. Mufc. p. 35. t. 6. f. 38. A. Raii Syn. p. 61.

On Rocks and Stones in the Sea, common.

---

IT is much to be regretted that Linnæus did not preferve fpecimens in his Herbarium of the few Confervæ which he has defcribed. From the fize and beauty of the prefent fpecies, added to its great abundance on every fhore, it appears almoft impoffible that it fhould have wholly efcaped his attention; but no defcription or reference is to be found in his works, which at all correfpond with it. The author who firft gave it the trivial name by which it is now generally known was Hudfon; he refers to the number above quoted of the Hiftoria Mufcorum, but Dillenius, as was conjectured in the Catalecta Botanica, has confounded two plants under that head; the firft of which, as appears by the Herbarium, is the prefent fpecies; the latter, according to the obfervations of my friends Dawfon Turner and Jofeph Woods, jun. is Fucus fubfufcus.

C. rubra often grows to the length of 18 or 20 inches, and varies from a dark to a light red or purple color, which is very liable to bleach. The root is a fmall callus, from which arife one or more filaments about the fize of fewing filk, and repeatedly divided without any regular order, though moft frequently in a dicho-tomous manner; the ramuli are fetaceous; the diffepiments of a dark red, and moftly more or lefs contracted; the joints beautifully reticulated, and pellucid towards the center. The capfules are feffile and lateral, more round than thofe of C. coccinea, but are precifely of the fame nature, as are alfo the feeds, except that when they iffue from the capfule, much lefs of the gelatinous pulp attends them. Each capfule is fubtended generally by one, but fometimes, as in my figure, by three fubulate ramuli, which I apprehend may be confidered as a kind of calyx; their nature I hope hereafter to be able further to elucidate.

It frequently happens that the joints in fome of the older fpecimens fwell, and thereby affume a more beaded appearance than in their ufual ftate. This has been, though erroneoufly, as is fhewn under C. diaphana, regarded as the C. nodulofa of Hudfon, and botanifts have puzzled themfelves in endeavouring to find fpecific diftinctions between the fame plant in different ftages of growth. Dr. Roth has erred in quoting as a fynonym of this fpecies, though with a mark of doubt, the C. fucoides of the Flora Anglica, which is extremely diffimilar.

It adheres but flightly to paper, and not at all to glafs.

A. C. rubra, natural fize.

B. Ditto magnified 3.

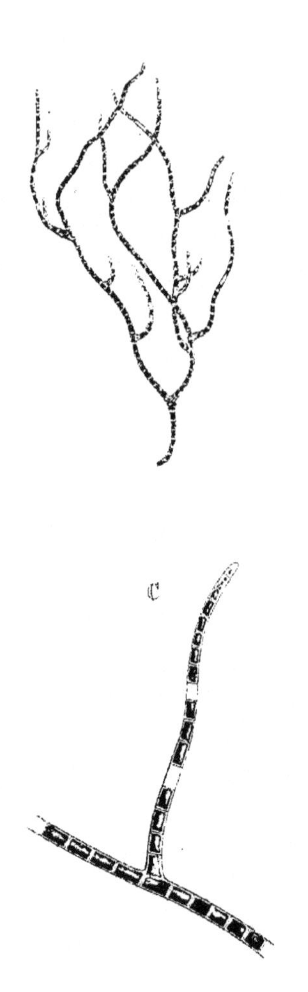


*Conferva aurea*

Published by L. W. Dillwyn Nov. 1. 1803.

F. Sansom sculp.

# CONFERVA AUREA.

C. filamentis ramofis aureis minutis; ramis longis patentibus rigidiuf. culis fub incurvis; diffepimentis pellucidis; articulis longiufculis.

Byffus aurea. Sp. Plant. p. 1638. Fl. Ang. p. 606. Fl. Scot. p. 1002. With. IV p. 144. Eng. Bot. t. 212.

Byffus petræa crocea glomerulis lanuginofis. Dill. Mufc. p. 8. t. 1. f. 16.

Byffus aureus Derbienfis humifufus. Raii Syn. p. 56.

In moift places, generally in a lime-ftone foil, frequently growing on Mufci: not very common.

---

WHATEVER claim the plants generally known by the name of Byffi have to be confidered a feparate genus, the prefent fpecies cannot juftly be ranked among them. Byffi are defined by Linnæus and fubfequent botanifts as confifting of fimple down or powder. Dr. Roth confiders them as folid fubftances with the feeds fcattered on the outfide, but this plant fo little correfponds with either, and has fo ftrikingly the ftructure of a Conferva, that I am furprized it has not been already referred to that tribe, inftead of being carried, as it has been by Dr. Acharius, to the Lichens, with which it has little affinity.

C. aurea occurs, though but rarely, in damp fituations on calcareous rocks, and in chalk pits, frequently forming irregular cufhion-like tufts on fome of the mufci; and when it grows in large patches, bears a ftriking refemblance, as Dr. Smith obferves, to a piece of orange-coloured cloth or velvet, and is a very confpicuous and beautiful object. Even without the aid of a microfcope, the filaments may be feen to be much branched; the branches are long, difpofed without any regular order, patent, moftly fomewhat incurved, and divided into

*'onferva villosa*

# CONFERVA VILLOSA.

C. filamentis ramofis; ramis ramulifque oppofitis diftantibus; articulis breviffimis; diffepimentis obfcuris villofis.

C. villofa. Fl. Ang. p. 603. With. IV. p. 141. Eng. Bot. t. 546.

On fubmarine Rocks and Stones. In Cornwall, *Hudson*. At Yarmouth, *Dawson Turner, Esq*. On the Rocks at the Mumbles, near Swanfea.

C. VILLOSA appears to have been firft obferved by the indefatigable author of the Flora Anglica, and may be reckoned among the moft unfrequent of this tribe, being found but in few parts of the kingdom, and not having been noticed by any foreign writer. Its growth feems to be very rapid, and its duration fhort, as it has, I believe, never been found but in the months of July and Auguft.

The whole plant is of a greenifh yellow color, and of a cartilaginous nature, but becomes foft and very flaccid foon after it is gathered. The root is a fmall callus. The ftem varies from fix inches to three feet or more in length; is confiderably thicker than horfe hair, and feldom more than thrice divided. The branches are diftant, moftly oppofite, and undivided when not more than two inches in length; the hairs, which conftitute the leading fpecific character, are difpofed in whirls on about every 4th or 5th joint, and moftly fubdivided in a fimilar manner, giving the plant a remarkably hairy appearance, as if befet with fome minute parafite; thefe hairs are extremely flender, and fo liable to be broken off, that it is almoft impoffible to find a fpecimen in which they are nearly all perfect. The diffepiments are difpofed at equal and very fhort diftances from each other; they are not readily difcoverable except in the verticillated hairs, to which when the juices have collapfed, as is moft commonly the cafe, they give a very beautiful appearance.

In drying, its color becomes more green, and it adheres to both glafs and paper.

A. C. villofa, natural fize.

B. Ditto magnified 2.

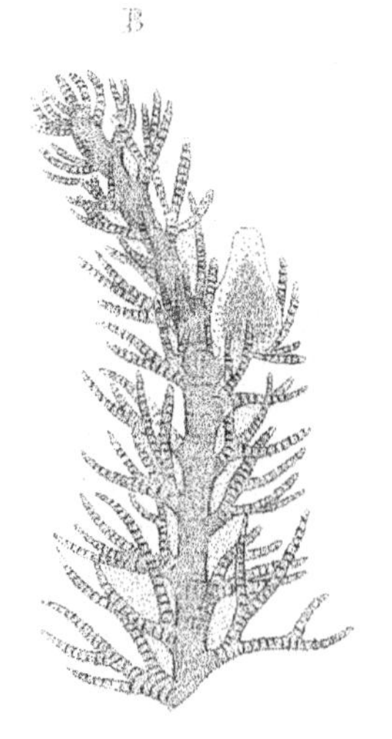


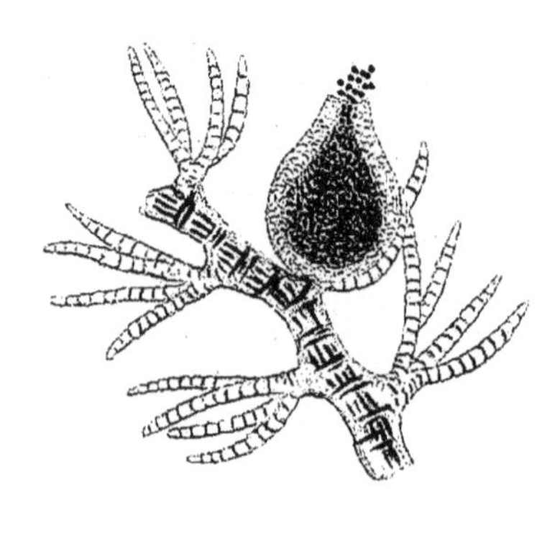

*Conferva coccinea.*

# CONFERVA COCCINEA.

C. filamentis ſub-cartilagineis ramoſiſſimis, hirſutis; ramis decompoſito-pinnatis; pinnis alternis; pinnulis ultimis faſciculatis pennicilliformibus; diſſepimentis obſcuris; articulis brevibus; capſulis ovatis.

C. coccinea. Fl. Ang. p. 603. With. IV. p. 140. Eng. Bot. t. 1055.

C. plumoſa. Ellis in Philoſophical Tranſactions LVII. p. 425. t. 18. f. c. c. d. D. Fl. Scot. p. 996.

Ceramium hirſutum. Roth, Cat. Bot. II. p. 169. t. 4.

Muſcus marinus purpureus parvus, foliis oblongis mille-folii fere diviſurâ. Raii Hiſt. p. 79. n. 25.

On Rocks and Stones in the Sea, common.

---

FEW marine productions exceed the preſent ſpecies in beauty or frequent occurrence, and none meets with more general admiration, or is more frequently gathered and uſed in ornamental devices by the female viſitors on our ſhores. The root is a ſmall callus, the frond ſolitary, the main ſtem nearly as thick as common twine, moſtly of a darker red than the branches, and of a more uneven and hairy ſurface. The primary ſhoots are diſpoſed without any regular order, of unequal lengths, and beautifully winged with alternate branches, which are pinnated with others, alſo alternate, and again divided into ramuli, iſſuing ſo nearly together as to give them a pencil-like appearance. The diſſepiments can ſcarcely be perceived in the main ſtem or primary branches, but are very apparent in the leſſer ones, and divide them into ſhort pellucid joints. The capſules, which are ſeſſile and of an oblong ovate form, appear in the Spring, in the earlieſt parts of which they are of a light red, becoming gradually darker, and in May the internal ſtructure, as repreſented at C. may be obſerved; the capſule, which is rather thick, contains a number of dark red ſeeds, immerſed in a clear gelatinous pulp, part of which iſſues with them, when ripe, through an aperture, formed by the burſting of the apex of the capſule. Not having been able to obſerve the diœcious fructification mentioned by Lightfoot, I am inclined to think that the plants which he ſuppoſed to be male and female, differed only in age. In June the capſules have generally ſhed their ſeeds, and during that month this plant is found lying in great abundance on the ſhore; this circumſtance may probably be accounted for by reflecting that the roots inſtead of inhering into the ſubſtances to which they adhere for the purpoſe of abſorbing nouriſhment, merely graſp the rocks for the ſake of ſupport, and it ſeems probable that when they have fructified, and their vigor begins to decline, they are no longer able to maintain their graſp, and therefore inſtead of decaying on their native ſpot, as is the caſe with land plants, eaſily yield to the preſſure of the tide, and are waſhed away to rot, or offer their ſervices to man on the ſhore. Several obſervations I made at Dover tend to ſtrengthen this poſition, which ſerves alſo to account for the ſudden diſappearance of the marine algæ mentioned in the Introduction to the Synopſis of the Britiſh Fuci, and confirmed by the experience of my friend the Rev. J. Lyons and numerous other marine botaniſts.

For reaſons given in my friend D. Turner's Synopſis above mentioned, p. 295. Ray's Synopſis cannot be here referred to. In drying, this plant undergoes but little change; it adheres to paper, but not at all to glaſs.

A. C. coccinea, natural ſize.
B. ditto magnified 3.
C. ditto ditto 2.

B

# CONFERVA DIAPHANA.

C. filamentis ramofiffimis; ramulis apice forcipatis; diffepimentis obfoletis; articulis utrinque torofis, medio pellucidis; capfulis fub-globofis lateralibus.

C. diaphana. Fl. Dan. t. 951. Fl. Scot. p. 996. With. IV. p. 139. Fl. Germ. III. pars. 1. p. 525. Cat. Bot. II. p. 226.

C. nodulofa. Fl. Ang. p. 600.

C. marina nodofa lubrica, ramofiffima, et elegantiffima rubens. Dill. Mufc. p. 35. t. 6. f. 38. A. Raii Syn. p. 62. t. 2. f. 3.

Rocks, Stones, and Fuci, in the Sea, frequent.

---

THIS Species, which is not an uncommon ornament of nearly every fhore, is in beauty furpaffed by few, prefenting to the naked eye the appearance of a feries of fmall beads alternately colored and pellucid. It varies from 2 to 6 inches in length, and in color through all the intermediate gradations between a reddifh brown and dark purple. The root, as in moft other marine fpecies, is a fmall callus, from which feveral bufhy filaments proceed; thefe are repeatedly branched; the branches difperfed without any regular order, but moft frequently dichotomous, and fubdivided into ramuli, which are forked at the apices; the forks approaching each other in a forceps-like manner, though not fo ftrikingly as in C. ciliata. The diffepiments are obfolete, but the joints are fwollen at each end, and of a deep red color, occafioned by reticulated veins, fimilar to thofe which cover the whole joint of C. rubra, but which, in this fpecies, leave the middle perfectly colorlefs and tranfparent. The capfules are nearly round, lateral, feffile, and often furrounded by 4 or 5 fhort incurved ramuli.

The following argument, nſed by my friend Dawſon Turner, to prove that C. nodulofa of the Flora Anglica ſhould be made a ſynonym of this ſpecies, appears to me ſo concluſive, that I have adopted it without heſitation. "It has been ſuppoſed that Mr. Lightfoot was the firſt botanical author who noticed this ſpecies. A ſuppoſition that ſeems juſtified from his making no reference to Dillenius, and from his C. diaphana being introduced as a new plant in the Appendix to the Flora Anglica. This idea is however very erroneous, for from the Dillenian Herbarium, in which good ſpecimens are preſerved, it is clear that this is the No. 40 of the Hiſtoria Muſcorum, and conſequently the C. nodulofa of Hudſon, by the admiſſion of which, a great deal of confuſion, with reſpect to references, is done away, and a plant that has always been conſidered one of the moſt doubtful among botaniſts is clearly eſtabliſhed." The ſpecimen correſponding with No. 41, to which Hudſon refers as his C. purpuraſcens, is a ſmall variety of this ſpecies, but Hudſon's deſcription is ſo ſhort that it will equally apply to many other ſpecies.

It adheres but ſlightly to either glaſs or paper.

A. C. diaphana, natural ſize.
B. Ditto, magnified 2.

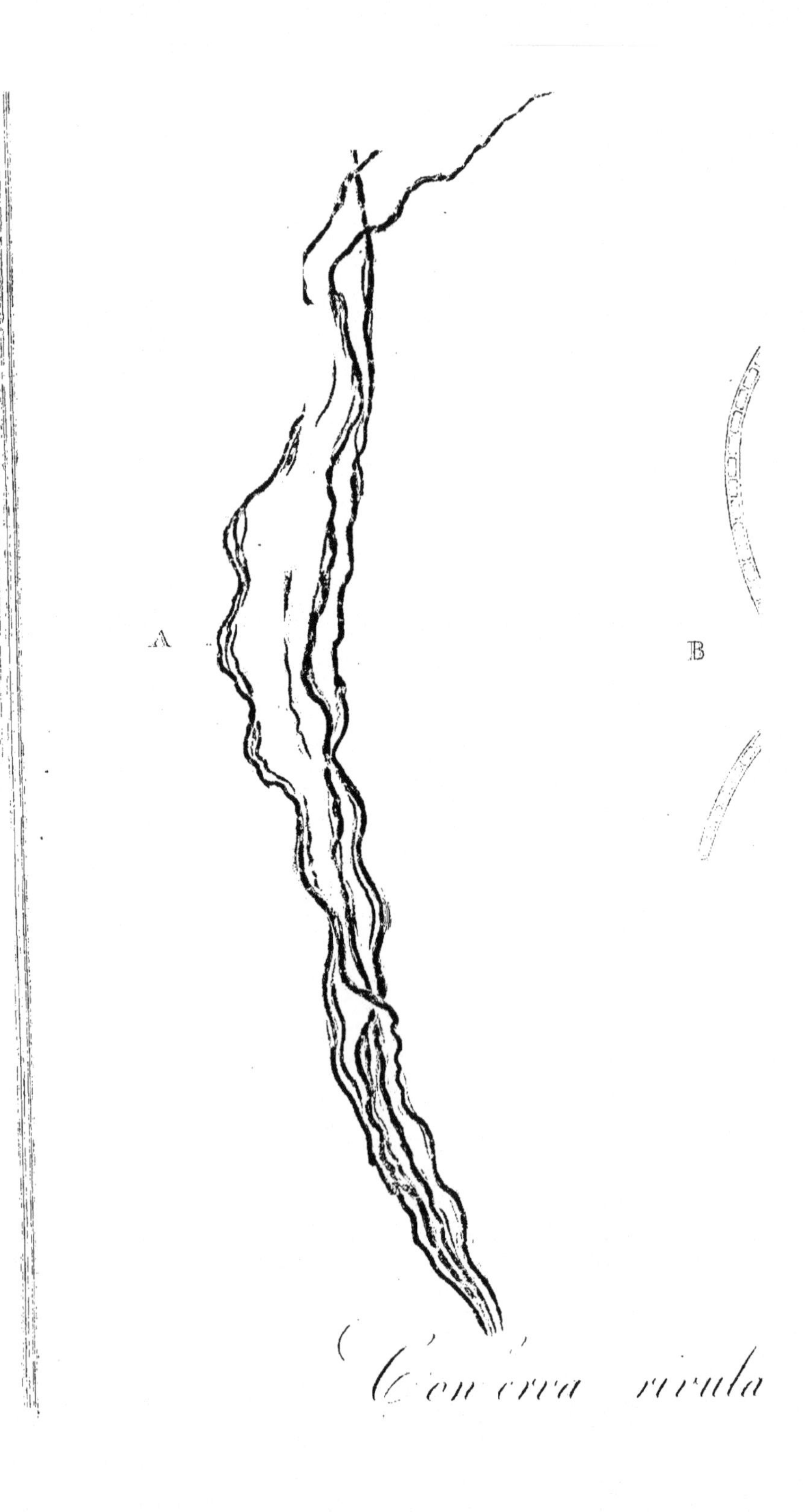


*Con erva rivula*

# CONFERVA RIVULARIS.

C. filamentis fimplicibus, atro-viridibus, tenuibus, longiffimis denfiffimè compactis, plerumque contortis; articulis breviufculis.

C. rivularis. S. Plant, p. 1633. Fl. Ang. p. 591. Fl. Scot. p. 975. With IV p. 127.

C. compacta. Roth. Fl. Germ. III. pars. 1. p. 497. Cat. Bot. I. p. 170.

C. fluviatilis fericea, vulgaris & fluitans. Dill. Muf. p. 12. t. 2. f. 1.

C. Plinii. Ray Syn. p. 58.

In flow ftreams and ditches—common.

---

AT a time when fo much doubt prevails among Botanifts, on what Dillenius intended by his C. fluviatilis fericea vulgaris & fluitans, which all other authors have referred to as their C. rivularis, it is with fome hefitation that I venture to publifh the prefent plant as that fpecies. The following reafons, however, appear to me fufficiently ftrong fully to juftify this ftep.—Firft. It is certain that the plant intended by Dillenius is fome very common unbranched fpecies, which grows to a remarkably great length, and has a filky appearance. In thefe particulars the prefent fpecies moft ftrikingly correfponds, which is not the cafe with any other plant.—Secondly. In the Dillenian Herbarium, which I have recently examined, the plant immediately referred to as his 't. 2. f. 1.' is a fpecimen of C. fpiralis, which from his defcription he appears to have confounded with it, attributing the difference of its appearance to its growth in ftagnant water.* There is alfo a fpecimen of the plant here figured, under the name of C. madrafpatana; and among the fynonyms and remarks on his C. fluviatilis fericea vulgaris & fluitans, we find the following: " Conf. madarafpatana, Allocopafhy Malaba-

* " In aquis vero ftagnantibus, quas quandoque intrat, brevior eft & lati expanfa cernitur." Dill. Mufc. p. 12.

rorum, Pluk, Almath. p. 63, quam meolim in ipſius Herbario ſicco vidiſſe memini, & cujus ſpecimen etiam habeo, non differt a vulgari hac."

I muſt allow that there are alſo two ſpecimens of the preſent plant in the Herbarium, under the name of C. paluſtris bombycina, to which ſubſequent authors have referred as their C. bulloſa; but neither Dillenius's figure or deſcription of that ſpecies agree at all with this plant, and this inaccuracy is by no means ſurpriſing when we reflect that Dillenius did not uſe a microſcope, and that his C. paluſtris bombycina contains all thoſe Confervæ, which generate and retain among their filaments a ſufficiency of air to raiſe them up, and enable them to float on the ſurface of the water, as is frequently the caſe with the preſent ſpecies.

C. rivularis grows in very compact ſilky ſlender maſſes, of a dark green color, frequently carried out to the length of two or three feet, and twiſted by the action of the ſtream. The filaments are ſimple and ſlender, of a uniform color, and divided into ſhort joints, which ſometimes appear filled with granules, that moſt probably are the fructification of the plant, no other having been diſcovered.

In very ſhady clear ſtreams I have ſeveral times found a plant approaching the preſent ſpecies in many particulars, but differing in being furniſhed with numerous ſhort ſpine-like branches, three or four of which moſtly iſſue from the ſame diſſepiment, and ſome being erect, and ſome reflected, preſent a curious appearance. Dr. Roth, and Profeſſor Mertens, in a letter to my friend Dawſon Turner, expreſs their opinion that it is but a variety of this ſpecies; but the above-mentioned, and ſome other more trifling differences, are ſo ſtriking, that with great deference to their experience in this tribe, I conceive that publiſhing them as diſtinct will be the moſt certain way to avoid all future confuſion.

The ancients attributed to it the power of uniting fractured bones, by binding it on the fracture, and keeping it conſtantly moiſtened with water. See Plin. Hiſt. Nat. Book 27. Chap. 9.

It adheres firmly to either glaſs or paper.

A. C. rivularis, natural ſize.

B. Ditto, magnified 1.

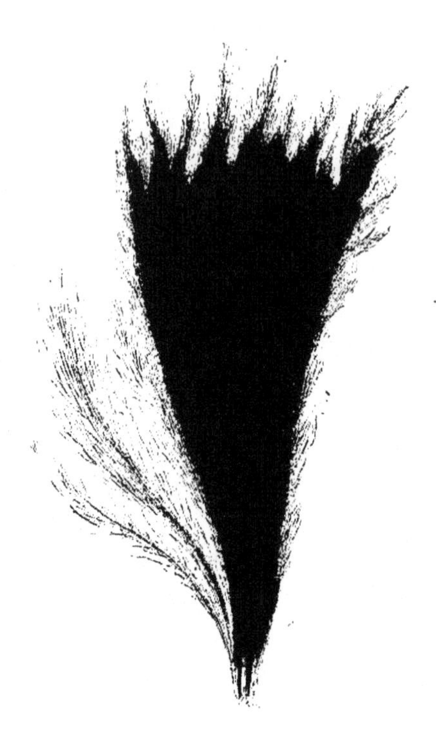

A

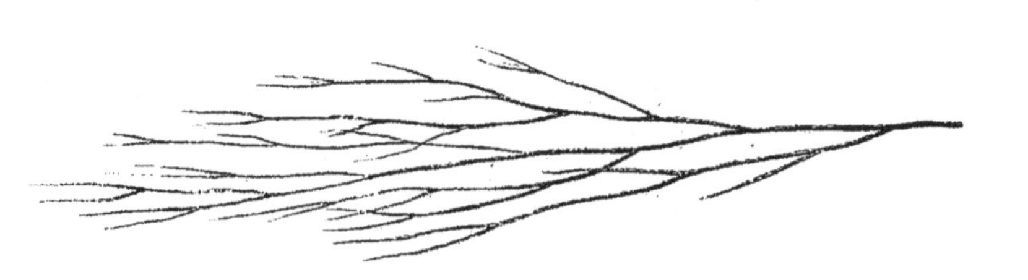

B

C. filamentis ſub dichotomis faſciculatis venoſis; articulis longis.

On rocks in the ſea at Dover and Swanſea.

---

THE firſt time I found this ſpecies was on the rocks near Archliff Fort, Dover, in 1799, but it had for many years before been gathered by M. Wigg, on the coaſt at Yarmouth, whence a ſpecimen of it was communicated by D. Turner to his learned friend Profeſſor Mertens, who gave it the name of C. ſtricta. It grows in thick bundles, ſeldom more than three inches in length, of a dull crimſon color. Many filaments riſe from the ſame root, in thickneſs about equal to the hair of the human head, and repeatedly divided and ſubdivided into branches and ramuli, for the moſt part alternate. Under the higher powers of the microſcope, the filaments appear as if compoſed of a number of longitudinal cylindric tubes, divided by dark diſſepiments at equal diſtances, and at the ſame part of the filament, and appearances make it highly probable that the filaments in this and ſome other marine ſpecies have no general diſſepiment, but that the tranſverſe line agreeing at firſt ſight with thoſe of C. glomerata, is in fact an aggregation of the diſſepiments of the before-mentioned cylindric tubes; and the tubes, eſpecially in the young and ultimate ramuli, are more or leſs ſpiral. The joints in length are about equal to thrice their thickneſs. There is no danger of confounding this with any other ſpecies; it approaches neareſt to C. ſetacea, but its more brilliant color, larger ſize, and far longer joints, at once diſtinguiſh that ſpecies.

MARINE BIOLOGICAL ... WOODS H... LIBRARY

In drying its color undergoes no change. It adheres firmly to paper, and ſlightly to glaſs.

A. C. ſtricta, natural ſize.

B. Ditto, magnified 4.

C. Ditto, ditto 1.

A

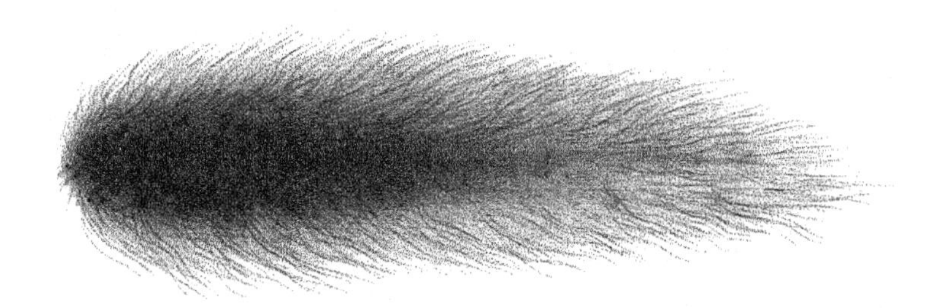

B

# CONFERVA AMPHIBIA.

C. filamentis ſub-articulatis, ramoſis denſiſſime implexis; ramis patentibus remotis; ramulis exſiccatione coeuntibus in aculeos; diſſepimentis parum contractis, capſulis ſeſſilibus, ſub ellipticis.

C. amphibia. Sp. Pl. p. 1634. Fl. Ang. p. 954. Fl. Scot. p. 979. Withering, IV. p. 129. Roth, Fl. Germ. III. p. 1. n. 7. Cat. Bot. I. p. 16. II. p. 192.

C. amphibia fibrilloſa & ſpongioſa. Dill. Hiſt. Muſc. p. 22. t. 4. fig. 17. B & C.

β. ramis elongatis.

C. furcata β. Fl. Ang. p. 592. Withering, IV. p. 128.

Ceramium cœſpitoſum. Roth. Fl. Germ. III. p. 1. p. 475. Cat. Bot. I. p. 154. II. p. 186.

Conferva paluſtris filamentis brevioribus & craſſioribus. Dill. Muſc. p. 17. t. 3. f. 10.

C. paluſtris ſub hirſuta filamentis brevioribus & craſſioribus. Ray. Syn. p. 447.

In ſmall pools and ſhallow waters. β. in ſtreams and deep waters.

---

AMONG flowering plants we find ſeveral inſtances of ſtriking varieties produced by the more or leſs watery ſituation in which individuals chance to grow, and perhaps no Botaniſt would acknowledge the two moſt oppoſite varieties of Myoſotis ſcorpioides, or Lotus corniculatus, to be the ſame ſpecies, without an opportunity of tracing them through their ſeveral gradations. The ſame may be ſaid of the preſent plant, which has hitherto formed two ſpecies, and it is only after a careful examination that I have here arranged them as one.

On the edges of ditches, and in ſimilar ſituations, it frequently occurs in maſſes, ſo denſely matted as to hold water like a ſponge, with its ſurface heſet by erect branches, which give it a very briſtly appearance. In this ſtate it is well known to Botaniſts as the C. amphibia of all modern authors. Its hue is a bright green, becoming aſh-colored with age. The root I have not been able to diſ-

A

B

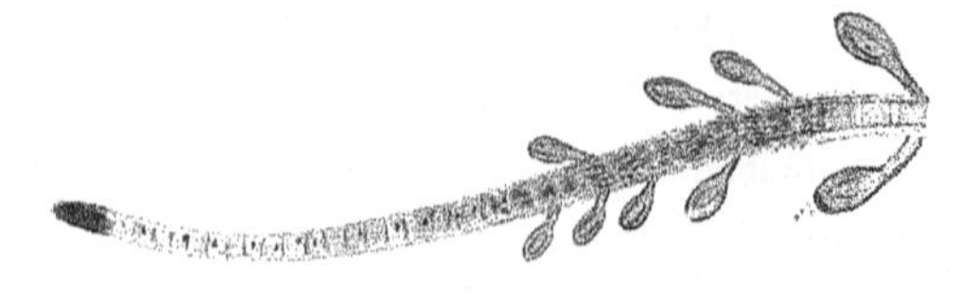

C

# CONFERVA SPONGIOSA.

C. filamentis ramofis; ramulis breviffimis fimplicibus undique imbricatis; articulis brevibus; capfulis oblongis pedicellatis.

C. fpongiofa. Fl. Ang. p. 596. Fl. Scot. p. 983. With. IV. p. 132. Roth, in Schrader's Journal, part II. 1800.

Fucus hirfutus. Lin. Mant. p. 134.

F. teretifolius fpongiofus pilofiffimus. Ray Syn. p. 46.

Mufcus marinus hirfutus; flagellis longis ramofis fub viridibus. Hift. Ox. III. p. 650. Sect. 15. l. 9. f. 6.

Rocks in the fea, not uncommon.

---

C. fpongiofa is not uncommon on our fhores, and is particularly abundant on the rocks at Cromer, Ilfracombe, and Swanfea. It feldom exceeds three inches in length, and varies from a very dark to a lighter olive color. The root is a callus, from which feveral irregularly branched ftems arife; the ftems and branches are clofely imbricated with fhort, fimple, rigid, hair-like ramuli, difpofed without any apparent order. In thefe ramuli fhort joints are readily difcoverable with the affiftance of a microfcope. The capfules are fmaller, and placed on longer footftalks than in any fpecies heretofore defcribed: they generally abound on the ramuli, and are frequently, though not conftantly, oppofite. The feeds are difcharged as defcribed under C. coccinea.

Relying on the rough and fpongy appearance of the prefent fpecies, and C. verticillata, Hudfon feems to have had no idea of feparating them, and we are indebted to the learned author of the Flora Scotica for firft afcertaining their difference. In the former the hair-like ramuli are ſimple, ſtrait, and diſpoſed without order; in the latter forked, incurved, and regularly verticillate.

In drying it changes to a darker color, and adheres to neither glaſs or paper.

A. C. ſpongioſa, natural ſize.

B. Ditto, magnified 4.

C. One of the ramuli of ditto, magnified 1.

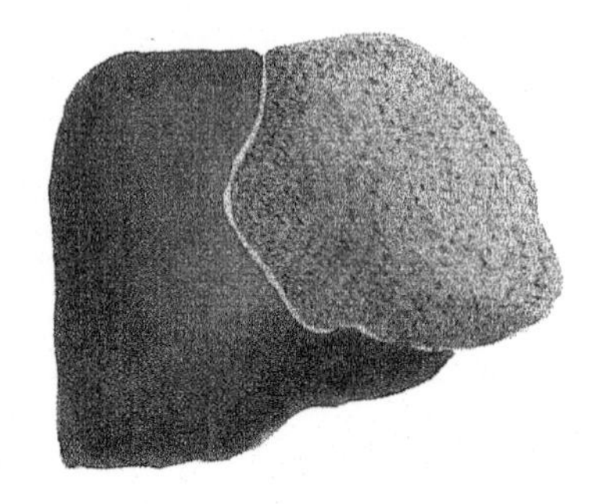

A

B

*Conferva purpurea*

# CONFERVA PURPUREA.

C. filamentis dichotomis flexilibus minutis; dichotomiis approximatis; diſſepimentis obſcuris; articulis longiuſculis.

Byſſus purpurea. Fl. Scot. p. 1000. Eng. Bot. t. 192. With III. p. 144.

Byſſus rubra. Fl. Ang. p. 605.

On rocks and ſtones, eſpecially ſuch as are near the Sea. Upon the baſe of the Abbot Mackinnons Tomb, in the ruined Abbey of Y. Columb-kill, *Lightfoot.* Near Aber, in Angleſea, *Rev. Hugh Davies.* In the Cavern under the Light-houſe on the Mumble rocks, and other ſimilar places near Swanſea.

---

THE ſtructure of the preſent ſpecies agrees ſo fully with that of C. aurea, that the reaſons already given for the introduction of the one among the Confervæ apply equally to the other, and need not therefore be here repeated.

Few of the minute productions of nature have a more elegant effect than C. purpurea. At the end of the cave above mentioned, it ſo entirely clothes ſome large rocks that they appear as if covered with the moſt beautiful crimſon or purple velvet; indeed the ſimilarity of appearance between this plant and velvet is wonderfully ſtriking, and far more ſo than in C. aurea, which may rather be ſaid to reſemble orange-colored or ſcarlet pluſh.

C. purpurea conſiſts of extremely ſhort flexile filaments, ſo denſely matted as to form an uniform maſs, reſembling the cruſt of a lichen. The color varies from a purpliſh crimſon to a darkiſh purple. Under the microſcope the filaments are ſeen to be repeatedly dichotomous at ſhort intervals, with dark colored diſſepiments, dividing them into joints, whoſe length conſiberably exceeds their thickneſs. I have not been able to find any fructification.

In dying it becomes crisp, and of a darker color than when fresh, and will not adhere to either glass or paper.

A. C. purpurea, natural size.

B. Ditto, magnified 1.

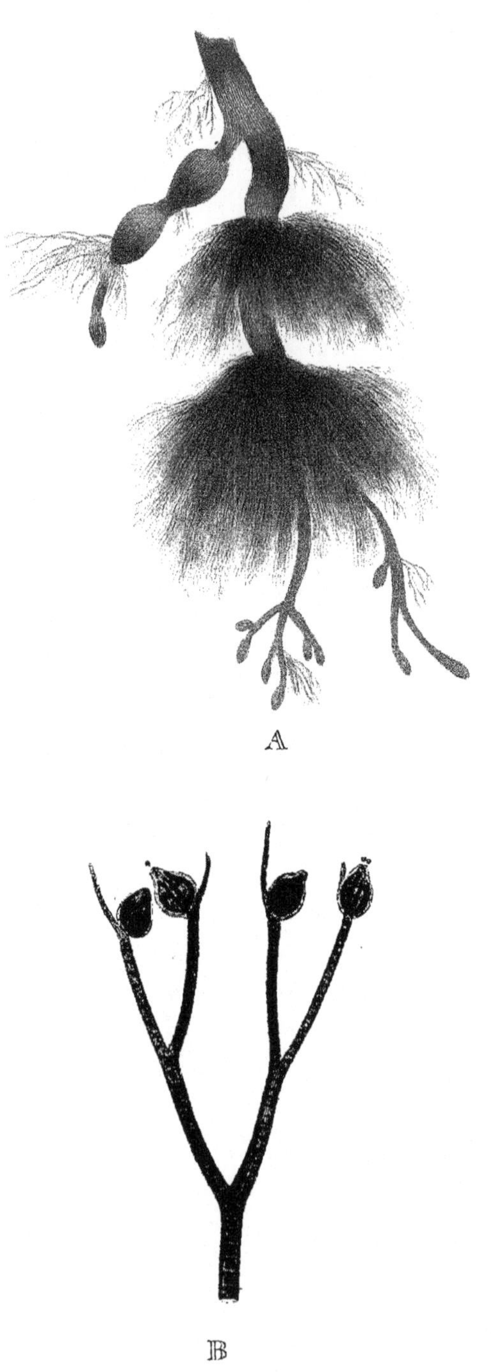
A
B

# CONFERVA POLYMORPHA.

C. filamentis dichotomis faſtigiatis ſub-cartilagineis, articulis brevibus, capſulis in ramulis ſuperioribus, ovatis, ſeſſilibus.

C. polymorpha. Sp. Pl. p. 1636. Fl. Ang. p. 599. Fl. Scot. p. 989. With III. p. 138. Ellis in Phil. Tranſ. LVII. p. 425. t. 18.

Ceramium faſtigiatum. Roth. Fl. Germ. III. pars. 1. p. 463. Cat. Bot. II. p. 175.

Conferva marina geniculata nigra palmata. Dill. Muſe. p. 32. t. 6. f. 35.

Conferva marina geniculata ramoſiſſima lubrica, brevibus & palmatim congeſtis ramulis. Ray Syn. p. 61.

In the Sea, on Fucus nodoſus, common.

---

NO Conferva is more common or has been longer or better underſtood than C. polymorpha. It grows paraſitically on ſome of the larger Fuci, but moſt commonly on Fucus nodoſus, forming thick tufts, about two or three inches in length. The color when young is a dark purple, but changes with age, or when dry, to black. The root is a callus, which is ſo ſmall, and in color ſo preciſely reſembles the Fuci to which it adheres, that it is difficult to diſtinguiſh it. It appears to me to throw out extremely ſhort creepers, the ends of which adhering to the rocks, become other Calli, and thus ſupply the bundles of filaments which always occur in this ſpecies. Two or more ſubcartilaginous filaments, of the thickneſs of horſe hair, generally riſe from the ſame root; the ſtems are repeatedly dichotomous, with rather acute angles, which cauſes the bundled appearance of the branches. The diſſepiments are black; the joints ſhort, and a black ſpot may frequently be obſerved in the middle, occaſioned by a partial collapſe of the juices. The capſules are diſpoſed on the ſides of the ultimate branches; before they burſt they are ovate, but afterwards contract towards their apices. In their younger ſtate, Ellis appears to have miſtaken them for male flowers.

In drying it undergoes no change, and adheres but ſlightly to either glaſs or paper.

A. C. polymorpha, growing on F. nodoſus, natural ſize.

B. Ditto, magnified 2.

A

B

# CONFERVA LANUGINOSA.

C. filamentis fub-fimplicibus minutiffimis, ferrugineis; articulis longiuf-culis, medio-pellucidis; capfulis feffilibus fecundis.

In the Sea, adhering to other Confervæ. At Swanfea, common.

---

THE filaments of C. rubra and fome other fpecies of Confervæ often affume a ragged appearance as if in decay: and it was with equal pleafure and furprize that I found this appearance occafioned by the prefent elegant parafite which is fo extremely minute that the higheft power of the microfcope is hardly fufficient to afcertain its ftructure. The filaments are fometimes fimple and fometimes branched, but I have never been able to find more than two branches on the fame filament: the capfules are round and feffile, and when two or more appear together, as is frequently the cafe, they are always difpofed on the fame fide of the filament.

It differs from C. cirrofa, for which alone it might be miftaken, in its much fmaller fize, ferrugineous colour, and pellucid joints. In fize it agrees with C. nana, but differs in almoft every other refpect: lanuginofa is moreover a marine, nana a frefh water fpecies.

In drying it adheres to either glafs or paper.

A. C. lanuginofa, growing on C. rubra, natural fize.
B. Ditto magnified 1.

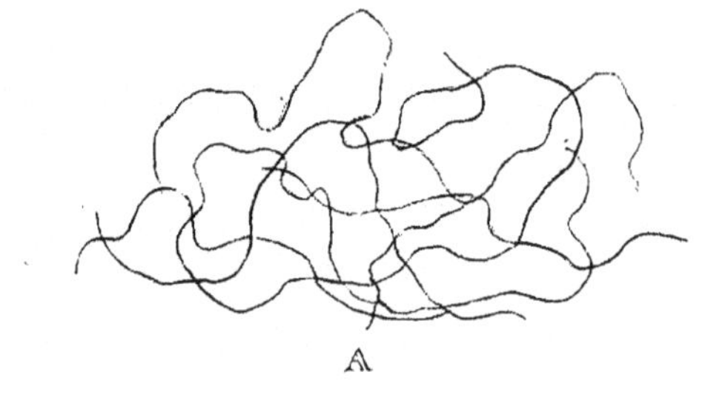


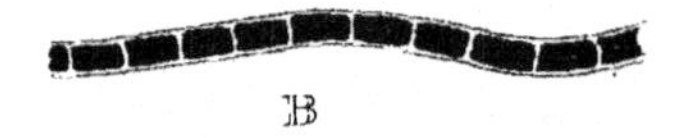


*Conferva tortuosa?*

# CONFERVA TORTUOSA.

C. filamentis ſimplicibus rigidiuſculis implicatis tenuibus; diſſepimentis pellucidis; articulis cylindraceis longiuſculis.

In Salt Pools by the River Yare near Yarmouth, and on Rocks in the Sea about Swansea.

---

THE preſent plant ſo nearly reſembles C. capillaris in miniature, and ſo well agrees with the moſt ſtriking characters of that ſpecies, that although it always appeared to me to be diſtinct, I heſitated on publiſhing it as ſuch till this opinion was confirmed by that of my friend Mr. Turner, and by Dr. Roth, and Profeſſor Mertens. I firſt found it in a Pool by the banks of the Yare, where C. capillaris alſo grew, and ſince on the rocks, and among the rejectamenta of the Sea at Swanſea. The filaments are as fine as the hair of the human head: their growth is curled and entangled as in C. capillaris, but not brittle or ſo rigid as in that ſpecies: it differs alſo in the joints which are nearly twice as long; nor have I ever obſerved the ſwelling of the diſſepiments already mentioned in the deſcription of that plant.

When taken out of the water and expoſed to air it becomes flaccid, and adheres but ſlightly to either glaſs or paper.

A. C. tortuoſa, natural ſize.

B. Ditto magnified 1.

A

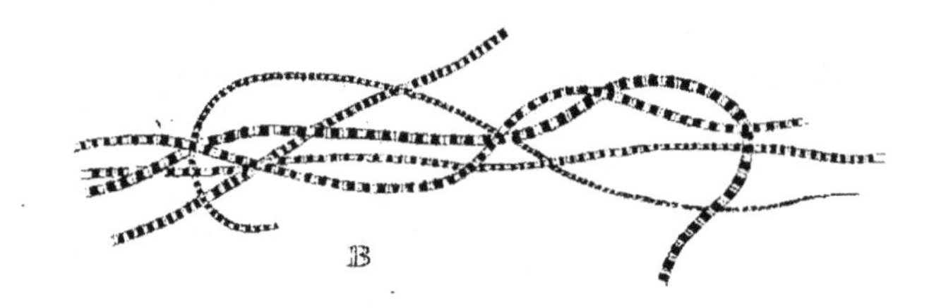


C

*Conferva lucens.*

# CONFERVA LUCENS.

C. filamentis ſimplicibus tenuibus glaucis lubricis; articulis breviuſculis; granulis in faſciis coacervatis.

On Rocks and Stones in clear rapid Rivulets. Frequent in Monmouthſhire. *Joſ. Woods, jun.* Alſo in Glamorganſhire.

---

THIS elegant ſpecies is found not unfrequently in the clear rapid rivulets of Glamorganſhire, and most probably of other mountainous Counties; but does not often occur in a perfect ſtate, the ends being extremely liable to be broken off, and the plant otherwiſe injured by the action of the current againſt the rocky and pebbly bottoms on which it grows, eſpecially when the ſtreams are ſwoln and flow with more than uſual rapidity.

The filaments are ſimple and ſlender, and taper towards the ends, in length ſeldom exceeding three inches. The joints are ſhort and almoſt pellucid near the diſſepiments with a band-like aggregation of granules in the middle. When gathered and placed in ſtagnant water the filaments greatly reſemble those of C. ſpiralis, but their different places and modes of growth will readily diſtinguiſh the two plants when growing. C. rivularis may at once be known from C. lucens by its darker colour, far greater length, and twiſted growth.

In drying it adheres to either glaſs or paper.

A. C. lucens, natural ſize.
B. Ditto magnified 3.
C. Ditto Ditto 1.

*Conferva læte virens.*

# CONFERVA LÆTE VIRENS.

C. filamentis ramofiffimis rigidiufculis arcuatis; ramulis alternatim fecundis; diffepimentis pellucidis; articulis longis.

Rocks, Fuci, and Corallines in the Sea. About Swanfea, frequent.

---

THIS fpecies is extremely common on the fhores of many parts of South Wales, but has not to my knowledge been obferved elfewhere. It grows indifferently on ftones, fuci, and corallines, and often nearly fills the bafons among the rocks, where it may at once be diftinguifhed from its congeners by its light green color and bufhy mode of growth. Its root, a small callus, gives rife to one, two, or more filaments which are from three to fix inches in length and irregularly branched; the branches are difpofed without much apparent order, fometimes dichotomous, or alternate, though not unfrequently three or four iffue fucceffively from the fame fide, they are much curved, and therein this fpecies differs ftrikingly from C. rupeftris, the branches of which are remarkable for their ftraitnefs: many together of the ultimate branches are arranged alternately on each lide of the fhoot, and thefe are again beset with ramuli difpofed in the fame order, of which one iffues from the end of nearly every joint. The diffepiments are pellucid and divide the filaments into joints whofe length varies very much in the fame fpecimen, but is always greateft in the principal filament and leaft in the ramuli. No fructification has been yet observed.

In drying, it preferves its colour, and adheres flightly to paper, but not to glafs.

A. C. lætè virens, natural fize.

B. Ditto magnified 2.

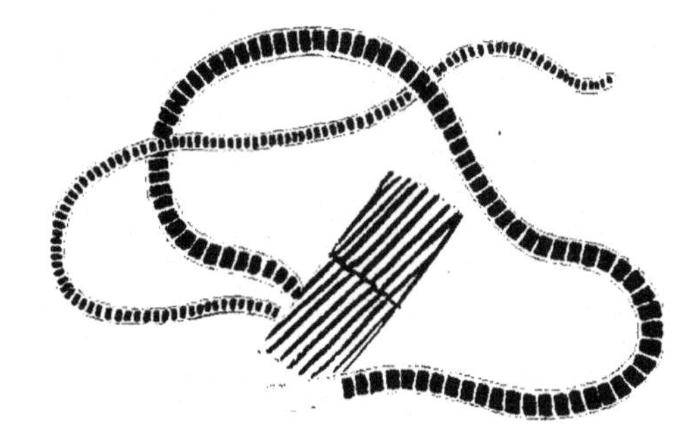

B

*Conferva flacca.*

# CONFERVA FLACCA.

C. filamentis ſimplicibus tenuiſſimis minutis flaccidis; diſſepimentis pellucidis; articulis breviſſimis.

In the Sea, adhering to Fuci and Confervæ.

---

THIS delicate paraſite has at preſent been only obſerved in the neighbourhood of Swanſea, but is moſt probably not uncommon elſewhere: it is found on Confervæ, on the ſmaller Fuci, and alſo ſometimes on the ſides of boats or other wood expoſed to the ſea water. It grows in looſe patches of a green colour, generally about half or three fourths of an inch in length. The filaments are almoſt univerſally, if not always, ſimple: among a great number which I have examined only one could be found with any appearance of ramification, and in this it is very possible I may have been deceived, as I could never find another. The diſſepiments are pellucid: the joints in length but little more than half their thickneſs. No fructification has been diſcovered. There is no chance of its being confounded with any other ſpecies with which I am acquainted: the much greater length of its filaments, and different mode of growth, will, at once, diſtinguiſh it from C. confervicola.

In drying it adheres firmly to glaſs and paper.

A. C. flacca, natural ſize.
B. Ditto magnified 1.

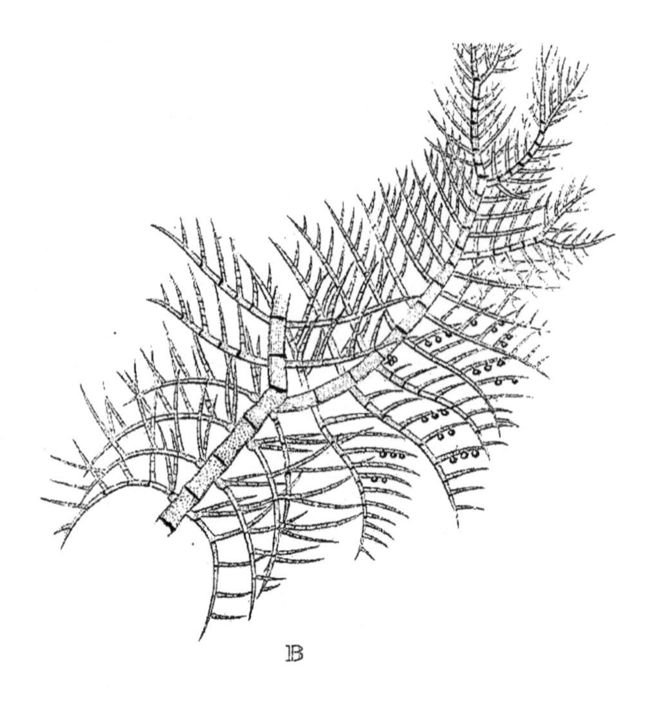

B

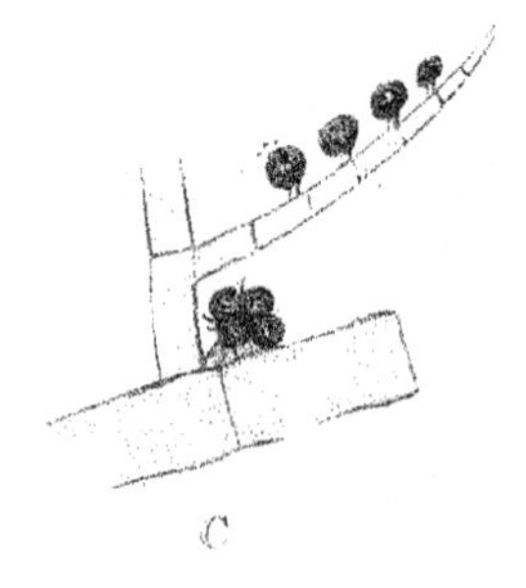

C

*Conferva plumula.*

J. Sowb.

# CONFERVA PLUMULA.

C. filamentis ramosissimis; ramis alternis pinnatis; pinnis oppositis; ramulis ultimis secundis; articulis longiusculis; capsulis brevius pedicillatis.

C. plumula. Ellis in Phil. Trans. LVII. p. 426.

In the sea, adhering to Confervæ. At Brighton, *Ellis.* In Caswell Bay near Swansea, during the summer months.

---

ELLIS gave an excellent drawing of this beautiful species to the Royal Society, in the year 1768, which was published by them in the 57th Volume of their Transactions; but since that time no Botanist appears to have noticed it, and it remained a desideratum till I met with it on the beach at Swansea, in August, 1802.

The plant is of a light red colour, and from the fineness of its filaments, has, when lying on the shore, the appearance of an Ulva in decay. The root I have not been able to observe, but we may fairly conclude, from analogy, that it is a minute callus. The whole frond is pellucid, with dark dissepiments: the branches are pinnate, with opposite pinnæ bearing smaller branches, arranged on one side only: the length of the joints varies very considerably, and is not unfrequently twice as great as in the annexed drawing. The capsules are very numerous, placed on short fruit stalks, arranged like the ultimate branches on which they grow; and, as in all other species allied to this, discharging their seeds by an orifice at the top. At C, I have given a sketch of four clustered together, which may occasionally be seen, but appears to be a lusus naturæ.

| | | | |
|---|---|---|---|
| A. | C. plumula, natural ſize. | | |
| B. | Ditto | magnified | 3. |
| C. | Ditto | Ditto | 1. |

A

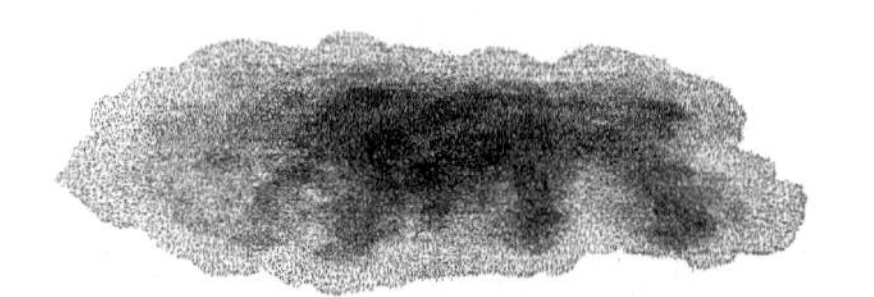

B

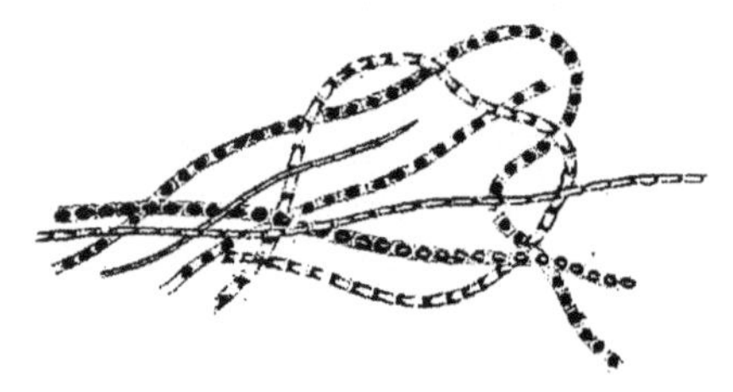

# CONFERVA PUNCTALIS.

C. filamentis fimplicibus lubricis tenuiffimis; diffepimentis obfcuris, articulis breviufculis cylindricis, fuccus in globulos folitarios demum congeftus.

C. punctalis, Muller in Nova Acta, Pet. III.

In Ditches and Pools not unfrequent.

---

PERHAPS no Botanift examining this fpecies in its younger ftate only would allow it to be the C. punctalis of Muller, nor till it has arrived at its maturity, when the green matter of the joints collapfes into a feries of globules; then, under any other than the higheft powers of the microfcope, the diffepiments by their extreme thinnefs entirely elude the clofeft obfervation, and the plant accords well with Muller's defcription, "filamentis inarticulatis, fimplicibus, ferie puncto-fum longitudinali."

C. punctalis is frequently met with in Pools and Ditches, as well on Heaths as in Marfhes. The colour of its filaments varies from a pale bright green to a yellowifh green; their length is from one to two inches; but what will at once diftinguifh this plant from all its congeners is their extreme tenuity, which is fuch that when fingle they can hardly be diftinguifhed by the naked eye: in this refpect C. punctalis refembles C. muralis, but the different color, place and mode of growth, and far different ftructure when examined with a glafs, preclude the poffibility of its being confounded with that fpecies. The diffepiments from their tenuity are obfervable only when a ftrong magnifying power is applied; they divide the filaments into joints, whofe length is about equal to their thick-nefs. Whilft the plant is young, thefe joints are nearly of a uniform greenifh color, but with age the green matter of each joint collapfes into a globule, and

7

occasions the aforementioned bead-like appearance; whether this is the fructification future observation must determine; no other has been discovered.

When dried it shines like C. pectinalis, and adheres firmly to both glass and paper.

A. C. punctalis, natural size.

B. Ditto, magnified 1.

A

B

*Conferva scoparia*

J. Simpki

# CONFERVA SCOPARIA.

C. filamentis ramofiffimis rigidis, ramis fafciculatis, ramulis alternis acuminatis, diffepimentis obfcuris, articulis brevibus.

C. fcoparia. Sp. Plant. p. 1635. Fl. Ang. p. 595. Fl. Scot. p. 981. With. IV. p. 131.

C. marina pennata. Dill. Mufc. p. 24. t. 4. f. 23.

Fucus fcoparia s. Pennachio marino. Bauhin pin. p. 366. Hift. III. 800.

On Rocks and Corallines in the Sea, not uncommon.

---

THE above references fufficiently prove that C. fcoparia is one of the few Confervæ which have been long known and well afcertained by Botanifts; indeed it is fo far from uncommon, is fo obvioufly different from every other fpecies, and with its cluftered branches often bears fo ftriking a refemblance to a painter's brufh or pencil, that it is almoft impoffible it fhould have been otherwife.

This fpecies when young is of a brownifh olive, changing with age to a ruffet brown. From a fmall callus one or more ftems arife, varying in length from two to fix, and Mr. Lightfoot fays to nine inches. The branches are numerous, alternate; the upper ones often fo much longer and more cluftered than the lower, as to give them a brufh-like appearance; they are every where befet with alternate fpine-like ramuli, which are highly characteriftic of the fpecies. The diffepiments are of a darker colour than the reft of the filament, and divide it into joints, whofe length about equals their thicknefs. No fructification has been difcovered. Its texture is remarkably like that of many Corallines, fo that doubts have arifen in the minds of feveral Botanifts how far it really belongs to the vegetable kingdom. Naked fpecimens of this plant are not unfrequently miftaken ſor Conferva pennata, which however is a very different ſpecies, ſeldom exceeding two inches, and formed of extremely thin, moſtly undivided filaments.

C. ſcoparia in drying will not adhere, or but very ſlightly, to either glaſs or paper.

A. C. ſcoparia, natural ſize.
B. Ditto magnified 3.

*Plate*

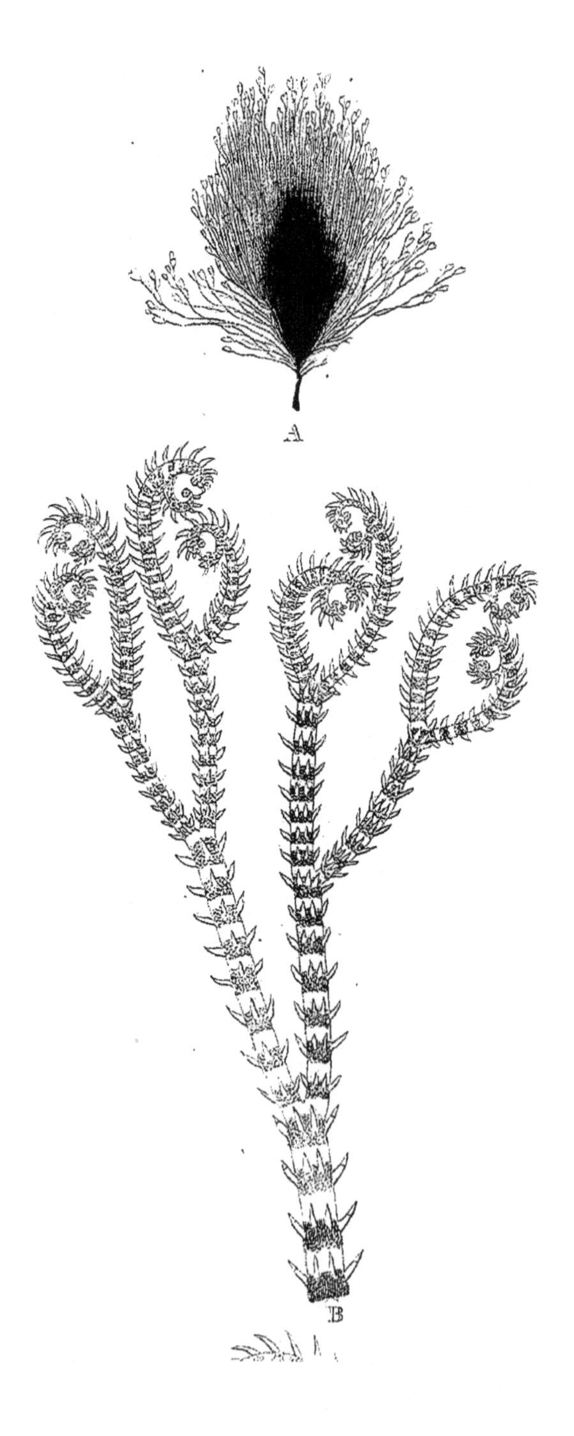

# CONFERVA CILIATA.

C. filamentis dichotomis apice forcipatis; diffepimentis verticillatim ciliatis articulis utrinque obfcuris medio pellucidis, capfulis fubglobofis lateralibus.

C. ciliata. Ellis in Phil. Tranf. LVII. p. 425. t. 18. f. h. H. H. Ang. p. 599. Fl. Scot. p. 998. With. IV. p. 137.

C. pilofa. Roth. Cat. Bot. II. p. 225. t. 5. f. 2.

Rocks, Stones, and Fuci in the Sea, not unfrequent.

---

THIS highly elegant Conferva, though fufficiently common on moft of our fhores does not appear to have been noticed by Linnæus, Ray, or by any author till Ellis publifhed an excellent figure of it in the 57th vol. of the Philofophical Tranfactions. It grows in bufhy maffes, feldom exceeding two inches in length, and varying in color from a bright to a purplifh red. The root appears to be a fmall Callus, from which feldom more than one ftem arifes, but I have fometimes obferved a connecting filament between thefe Calli, which whether it fhould be confidered as a creeping ftem or root I am at a lofs to decide, not having been able to feparate it from the fubftance on which it grows. The filaments are branched; the branches repeatedly dichotomous, remarkably incurved at their extremities in a forceps-like manner. The diffepiments are obfolete, but the joints at each end are generally more or lefs fwollen, and of a reddifh color, occafioned by reticulated veins, which as in C. diaphana leave the middle of the joint perfectly colourlefs and tranfparent; what however ftrikingly diftinguifh this from that fpecies are whirls of pellucid fpines which encircle each diffepiment, and give this plant a beautiful appearance under the microfcope. The capfules are roundiſh, lateral, nearly ſeſſile, and moſtly accompanied by three or four ſhort incurved ramuli.

It adheres ſlightly to paper, but ſcarcely at all to glaſs.

A. C. ciliata, natural ſize.
B. Ditto, magnified 3.
C. Ditto, ditto 1.

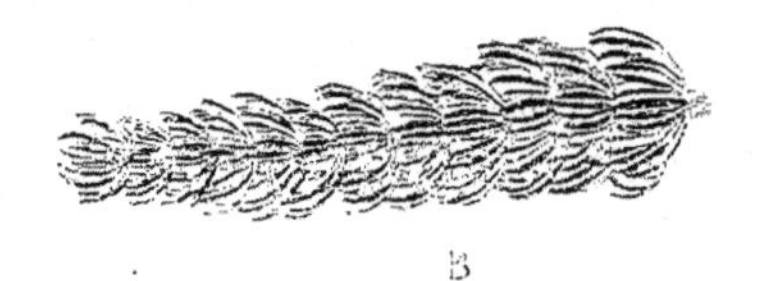


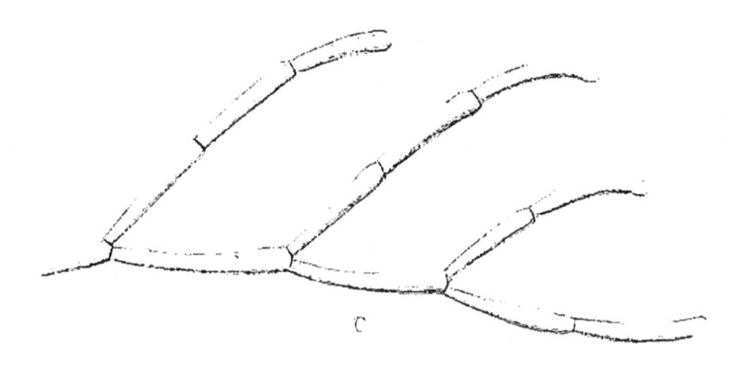


*Conferva equisetifolia.*

# CONFERVA EQUISETIFOLIA.

C. filamentis ramoſſiſſimis; ramis acuminatis elongatis ſubſimplicibus; ramulis verticillatis brevibus dichotomis articulis ramulorum longis.

C. equiſetifolia. Fl. Scot. p. 984. With. IV. p. 133.

C. multifida. Fl. Ang. p. 596. With. IV. p. 132.

C. imbricata. Fl. Ang. p. 603. Roth, Cat. Bot. I. p. 189?

*C. verticillata.* Roth in Schrader's Journal, III. p. 332? Schmidel Iter. t. 2.

Muſcus marinus hirſutus, flagellis longuribus rarius diviſus ruber. Hiſt. Ox. III. p. 650. f. 15. t. 9. f. 7.

On Rocks and Stones in the Sea, not unfrequent.

---

C. Verticillata and C. Spongioſa, already figured in this work, are very nearly allied to the preſent ſpecies; its red clay color and acuminated branches, with the conſtantly dichotomous ramuli, and their long joints, will however readily diſtinguiſh it.

It is occaſionally found on moſt of the Britiſh Coaſts, generally more or leſs covered with other Confervæ growing paraſitically on it. The length is from five to eight inches, the thickneſs nearly that of a crow's quill, and the color a dull red; the ſtem ſolitary and repeatedly branched; the branches are ſubulate and vary much in their diſpoſition; in ſome ſpecimens they are numerous, ſhort, and branched, in others long and nearly ſimple: in the ſame plant they vary alſo, as, though their diſpoſition is moſtly alternate; ſeveral together not unfrequently iſſue from the ſame ſide of the ſtem. The ſtem and branches are every where clothed with a profuſion of ramuli iſſuing in whirls from the end of each joint, which being longer than the joints are tiled on each other and give the plant a very rough and ſpongy appearance. In the ſtem and branches the joints are cylindrical and ſhort, but in the whirled ramuli the length is generally four or five times greater than the thickneſs, and they are ſlightly contracted at the lower and thickened at the upper end. On the older branches, particularly about the root, they are frequently ſwollen, and aſſume more or leſs of a globular appearance, in the ſame manner but more ſtrikingly ſo than in C. littoralis. The fructification has not yet been diſcovered.

The figure given by Schmidel of this plant in his journey, above quoted, is ſo excellent, that it is hardly poſſible it ſhould be miſtaken; but though Dr. Roth refers to this figure, and even ſays that his ſpecimen comes from Schmidel's her barium, yet as he deſcribes the whirled branches as conſtantly ſimple, I have thought it right to quote him with a mark of doubt. I have referred C. multifida of Hudſon, as well as his C. imbricata, to this ſpecies, on the authority of an authentic ſpecimen communicated by the Rev. Dr. Goodenough to D. Turner.

In drying the juices collapſe into a red parenchymous line, and it adheres to neither glaſs nor paper.

A. C. equiſetifioli, natural ſize.

B. Ditto magnified 6.

C. One of the whirled ramuli, ditto 2.

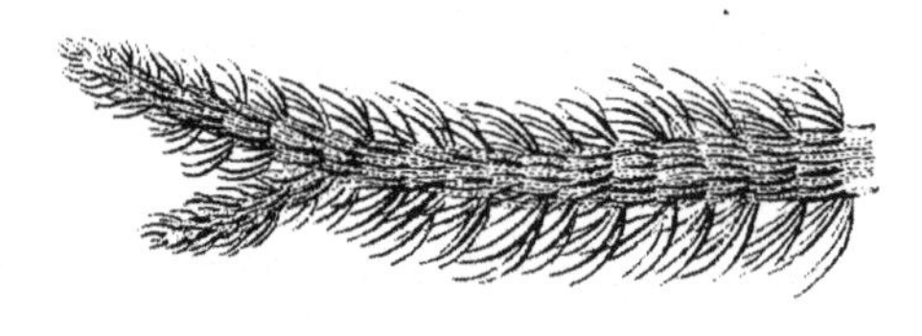

J. Sowerby sc.

*Conferva verticillata.*

# CONFERVA VERTICILLATA.

C. filamentis cartilagineis subdichotomis; ramulis ad dissepimenta verticillatis brevissimis incurvatis plerumq. bifurcis; articulis brevibus.

C. *verticillata.* Fl. Scot, p. 984. With. IV. p. 133.

C. *myriophyllum.* Roth in Schrader's Journal, III. p. 335.

On Rocks and Stones  the Sea, not unfrequent.

---

C. Verticillata is extremely plentiful in the pools left by the tide about Dover, and is more or less frequently met with on most of our coasts. It is generally from four to five inches in length, and of a dull olive color: the root is a callus from which several irregularly branched filaments arise: the stem and branches are of a horny nature, and every where beset with close whirls of rigid incurved, hair-like ramuli, which are mostly forked but sometimes simple, and though short yet twice as long as the joints of the stem. In these ramuli short joints are faintly observable with a microscope, very nearly resembling those of C. spongiosa, to which this plant is closely allied, but from which it may in general be at once distinguished by its forked, incured and regularly verticillate ramuli; but specimens sometimes occur so intermediate that it is not easy to determine to which they belong. The fructification has not been discovered, but is most probably similar to that of C. spongiosa.

It has already been remarked in the description of C. spongiosa, that the present plant was confounded by Hudson with that species, and we are indebted to Lightfoot for having first separated them. Roth, in the first volume of his Catalacta Botanica, and in Schrader's Journal, has followed Schmidel, and described C. equisetifolia under the name of C. verticillata, though he has erred in ascribing to it simple ramuli.

In dying it will not adhere to glaſs or paper.

| | | | |
|---|---|---|---|
| A. | C. verticillata, | natural ſize. | |
| B. | Ditto | magnified | 3. |
| C. | Ditto | ditto | 1. |

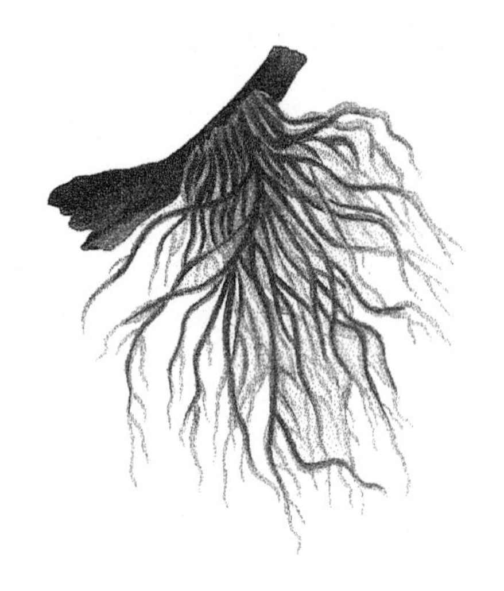

# CONFERVA TOMENTOSA.

C. filamentis ramofiffimis tenuiffimis denfiffimè implexis ramis divaricatis ultimis fimplicibus articulis longis.

C. tomentofa. Fl. Ang. p. 594. Fl. Scot. p. 982. With. IV. p. 130.

C. ceramium tomentofum β. Roth. Cat. Bot. II. p. 181. Fl. Germ. III. pars. 1. p. 470.

Conferva marina tomentofa, minus tenerea et ferruginea. Ray, Syn. p. 59. Dill. Mufe. p. 19. t. 3. f. 13.

In the Sea frequent, generally growing on Fucus veficulofus.

---

ALTHOUGH, according to the remarks of my friend Dawfon Turner in the feventh volume of the Linnean Tranfactions, the fpecimen preferved in the Dillenian Herbarium as *Conferva marina, tomentofa, &c.* is only a bad fpecimen of *C. littoralis*, yet the defcription in the Hiftoria Mufcorum, and alfo the original drawing in Sir Jofeph Banks's Library, feem to prove that Hudfon was correct in referring that fynonyma to the prefent fpecies.

The color of C. tomentofa is a pale greenifh or ruffet brown, remarkably deftitute of glofs, efpecially when dried: the length generally from three to five inches. Its filaments are repeatedly branched, fo extremely flender as to hardly be difcernible without a microfcope, and fo entangled and twifted together in rope-like coils as to make it abfolutely impoffible to feparate without breaking them. The branches iffue nearly at right angles; about the root they are rather numerous, but lefs fo towards the end, and the terminal ones are long and fimple; the length of the joints is at leaft three times as great as their thicknefs; they are perfectly cylindrical, and when examined under a glafs generally appear quite colourlefs, hut the diffepiments are dark brown. Dr. Roth defcribes the fruit of the plant as confifting of fcattered globular feffile capfules, but thefe I have not feen myfelf, nor am I aware that they have been found in England. How far that learned author is right in making *Conf. albida* Huds. a variety of *C. tomentofa*, is what I have yet no means of determining. The neareft affinity of the prefent fpecies is *C. littoralis*, with which it is fo frequently confounded, that though by no means an uncommon fpecies, it is one of thofe which are leaft accurately known to Britifh botanifts, it may however at once be diftinguifhed from that plant by its paler color, its diffimilar mode of growth, its different ramification, and long joints.

In dying it retains its color, and adheres though not firmly to both glafs and paper: the filaments in this ftate are ftill more clofely matted than when frefh, fo that the plant has the appearance of being nearly allied to C. fpongiofa, or by a young botanift may even be miftaken for that fpecies.

A. C. tomentofa, natural fize.
B. Ditto magnified 4.
C. Ditto Ditto 1.

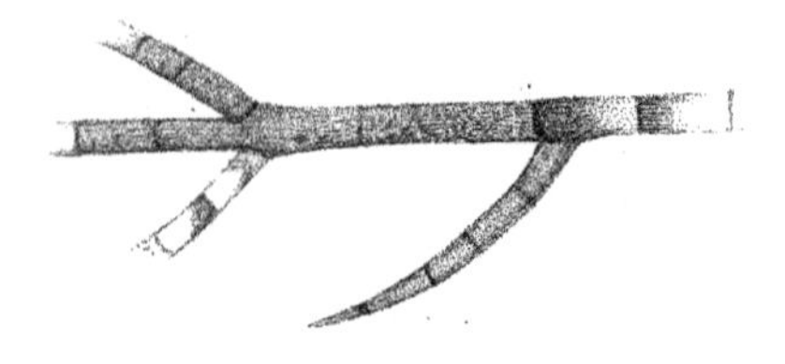

# CONFERVA LUBRICA.

C. filamentis ramoſiſſimis tenuibus longiſſimis ſplendenter lubricis rámis aculeiformibus articulis breviuſculis.

In clear Rivulets on Stones and Wood. At Lounde, near Yarmouth; and on Sketty Burrows, near Swanſea.

---

THIS elegant Conferva, which I firſt found ſparingly in a rivulet at Lound, near Yarmouth, where my friend D. Turner and myſelf have ſince repeatedly looked for it in vain, abounds in a clear ſtream on Sketty Burrows, near Swanſea. I cannot find that it has been heretofore deſcribed, and there is no other ſpecies to which it can be referred, or with which it can poſſibly be confounded.

It grows on wood or ſtones in large gelatinous maſſes, frequently from ſix inches to near a foot in length. The color is green with a ſlight tinge of blue; the filaments are very much branched; the branches diſpoſed without any apparent order, but uniformly iſſuing at an acute angle with the ſtem. The ultimate ramuli are numerous, moſtly ſhort, thornlike, and diſpoſed alſo without order, being ſometimes alternate and three or more not unfrequently iſſue together from the ſame joint. The joints are ſhort, and when the juices have collapſed, as is generally more or leſs the caſe, they give the plant a very beautiful appearance under the microſcope. The fructification has not been diſcovered.

In drying it adheres to both Glaſs and Paper.

A. C. lubrica, natural ſize.
B. Ditto, magnified 3.
C. Ditto, ditto 1.

*Conferva byssoides.*

# CONFERVA BYSSOIDES.

C. filamentis decompoſito-pinnatis; ramis ramuliſq; alternis, extremis perbrevibus, ſubfaſciculatis; diſſepimentis ex venarum anaſtomoſibus; articulis longiuſculis, capſulis ovatis ſeſſilibus.

*C. byſſoides.* Eng. Bot. p. 547.

*Fucus byſſoides.* Goodenough and Woodward, in Linn. Trans. III. p. 229.

On Rocks, Stones, and Fuci in the Sea, common.

---

THE preſent ſpecies was firſt deſcribed by Dr. Goodenough and Mr. Woodward, under the name of Fucus byſſoides, in the Tranſactions of the Linnean Society. It was not however without conſiderable heſitation that they thus arranged it with the Fuci; and Dr. Smith ſoon after in Engliſh Botany, removed it to the Confervæ, to which it properly belongs, as its congeners are at preſent all placed in this genus, and among them are ſeveral whoſe diſſepiments have an equally ſmall appearance of being formed by 'annular ſtrictures.'

C. byſſoides is extremely common on moſt of our ſhores; it grows in large maſſes, varying in length from three to ten inches, and in color from a reddiſh-brown to a light or purpliſh red. The root is a minute callus. The filaments are triply or quadruply pinnated, extremely flaccid, flexuoſe, pellucid and beautifully ſtriated by longitudinal veins, each of which arching over at or near the ſame place appears to form the diſſepiment. The branches and ramuli are all alternate; the primary branches long, the extreme ones very ſhort, and ſubfaſciculate; giving the plant throughout a ſingularly tufted appearance. The joints are rather long; capſules ovate, ſeſſile, moſtly axillary, reticulate, and preciſely of the ſame nature as thoſe of C. coccinea.

In drying the color becomes a dark dull brown, foon changing almoft to black, and it adheres, though but flightly, to either Glafs or Paper.

A. C. byffoides, natural fize.

B. ditto, magnified 2.

C. Capfule of ditto ditto 1.

B
C

# CONFERVA VIVIPARA.

C. filamentis dichotomo-ramofis, ramis flexuofis ad diffepimenta bulbiferis, bulbis piliferis, articulis longis, capfulis lateralibus feffilibus.

In boggy rivulets, growing on ftones and mofs, &c. near Yarmouth. *Dawfon Turner, Efq.* Near Cadoxton juxta Neath, Glamorganfhire. *W. W. Young.* On a heath about a mile weft of Five Lanes, between Launcefton and Bodmin, Cornwall.

---

THIS moft interefting fpecies I received at the latter end of May, 1802, from my friend Dawfon Turner, who firft difcovered it in the neighbourhood of Yarmouth, fince which it has been once found, though in fmall quantities, near Neath; and in September laft I was fo fortunate as to meet with it on a boggy heath in Cornwall, where in feveral rivulets it almoft clothed large maffes of Sphagnum latifolium. Mr. Turner has alfo received it under the name of C. pumilio, from Profeffor Mertens, who gathered it near Bremen.

It grows on various fubftances, in fmall, delicate, bufhy tufts, never I believe exceeding half-an-inch, while its ufual length is not more than two lines. The color is a yellowifh green, affuming a browner tinge with age. The ftem is irregularly dichotomous, and flexuous, as alfo are the branches, and under the microfcope they have rather a woody appearance. The length of the joints is about five times greater than their thicknefs. The fructification is in feffile capfules, at the end of the joints. At moft of the diffepiments where there are no capfules, a fmall bulb or bud is obfervable, from which proceeds a very long, unbranched, extremely flender, colorlefs filament, fimilar to the hairs of the Rivulariæ, and jointed, but with joints far longer than thofe of any other part of the plant. Thefe fmall bulbs iffue only from the ends of fuch joints as produce no capfules, and they appear to me precifely to correfpond in nature with the viviparous bulbs in feveral phænogamous plants, and their long filaments have greatly the appearance of being occafioned by the premature vegetation of their germs.

For the magnified drawing I am indebted to my friend Jofeph Woods, jun. Efq.

In drying it changes to more of a dull afh-color, and adheres to both glafs and paper.

A. C. vivipara, growing on Sphagnum latifolium, natural fize.
B. ditto, magnified 2.
C. ditto, ditto 1.

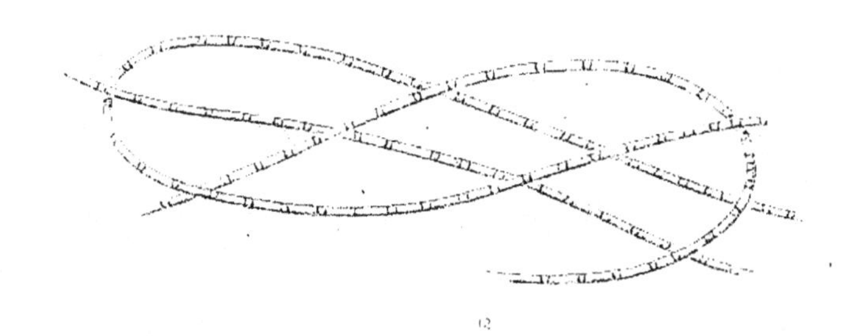

*Conferva sordida*

# CONFERVA SORDIDA.

C. filamentis fimplicibus tenuibus, diffepimentis annularibus, articulis longiufculis pellucidis.

**C.** fordida. Roth, Cat. Bot. I. p. 177. t. 2. f. 4. Fl. Germ. III. part. 1. p. 504.

In Ponds, Pools, and Ditches frequent.

---

SOME time fince I fent fpecimens of this plant marked C. fordida, together with the magnified drawing reprefented at B to Dr. Roth, requefting his opinion, and he favored me with the following remark: "Omni modo convenit cum mea Conferva fordida; at genicula parum contracta in meis fpeciminibus, quod forfan ab aetatis diverfitate dependet."

C. fordida in pools where the water has long remained without much motion, forms round the grafs or reed on which it grows a femi-tranfparent cloud-like mafs, of a yellowifh green color, but this readily yields to a fmall current, and the plant then floats in denfer maffes on the furface. When the water has been turbid, thefe maffes become mottled by the finer parts of the decayed vegetable matter and mud which lodges on them, and they then affume a dirty appearance. The filaments are very long, but it is difficult to afcertain the precife length from their entangled mode of growth; they are fimple and extremely flender; the diffepiments from the cylindricity and tranfparency of the filament appear like rings, and in fact thefe rings only are apparent, and it is only from analogy that I have fuppofed a diffepiment to exift, and from the probability that it may be tranfparent in common with the other parts of the filament. The length of the joints is moftly about equal to four times their diameter; they are frequently perfectly colorlefs under the higher powers of the microfcope, but a flight tinge of green is then obfervable in young and perfect fpecimens. No fructification has been difcovered.

From C. rivularis and C. genuflexa in a young ſtate with which alone of the ſpecies heretofore deſcribed there is any danger of confounding it, it may be diſtinguiſhed by its lighter color and pellucid joints. From the former it alſo differs in its mode and place of growth, its longer joints, and in the greater tenuity of its filaments, and from the latter in its much greater length.

In drying C. ſordida adheres to both Glaſs and Paper.

A. C. ſordida, natural ſize.

B. Ditto, magnified 1.

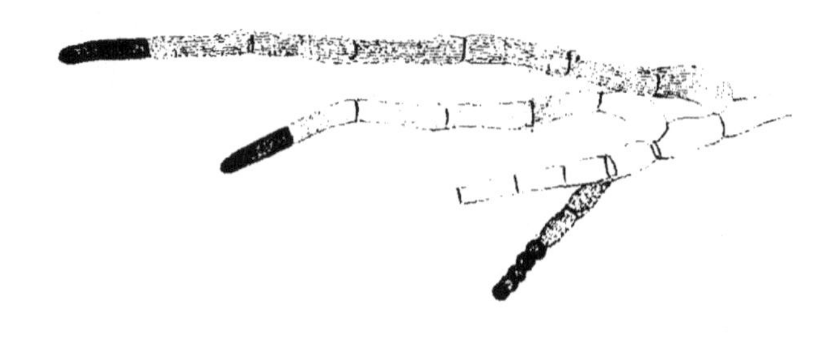

*Conferva umbroni*

# CONFERVA UMBROSA.

C. filamentis ramofis repentibus fragilibus brevibus-obtufis, ramis curvatis fimplicibus fubfecundis articulis longis cylindraceis inflatifque.

C. umbrofa, Roth. Cat. Bot. I. p. 191. t. 4. f. 3. Fl. Germ. III. part. 1. p. 521.

C. Arenaria. Roth. Cat. Bot. II. p. 217.?

On Boggy Ground near Swanfea.

---

THERE is every reafon to believe that the prefent fpecies is fufficiently common in certain fituations, though I am not aware of its having been noticed in Britain, till I lately detected it on part of a bog, the furface of which had been recently burnt, adjoining Singleton Wood, near Swanfea: probably however it has been often paffed by as a variety of C. frigida, to which, till placed under a microfcope, it bears a great refemblance, but may be diftinguifhed by its growing in fmaller patches, and by its darker color. It was firft difcovered by Dr. Roth, who figured and defcribed it in the firft Fafciculus of his Catalecta Botanica, under the name of C. umbrofa, and who, in the fecond Fafciculus, has given another fpecies, under the name of C. arenaria, which I apprehend is only a variety of the fame, as I have feen the joints fhort and inflated in one branch, whilft in another on the fame plant they were cylindrical, and in length fully equal to fix times their diameter.

The filaments are creeping, and fo remarkably fragile, that it is difficult to afcertain their length, which I believe never exceeds and feldom attains to half an inch. There are generally four or five branches which are fimple, and moft frequently difpofed on the fame fide of the ftem, but fometimes alternately; the apices are every where blunt: the diffepiments are more or lefs contracted and divide the filaments into joints, which vary greatly in fhape and length, as before obſerved ; and that which forms the apex of the branch is often of a darker color than the others. The fructification has not been diſcovered.

In drying it adheres to glaſs, and aſſumes though in a much leſs degree, ſomewhat of that ſhining appearance which is ſo ſtriking in C. pectinalis.

A. C. umbroſa, natural ſize.
B. ditto, magnified 3.
C. ditto, ditto 1.

# CONFERVA OCHRACEA.

C. filamentis ramofiffimis tenuiffimis, perfragilibus denfiffimè compactis, gelatinam ochraceam tamen in floccos fecedentem conftituentibus.

C. ochracea. Roth, Cat. Bot. I. p. 165. t. 5. f. 2. Fl. Germ. p. 494.
In Pools and Ditches, common.

---

THIS fingular fpecies is far from uncommon in Pools and Ditches, more efpecially in boggy fituations, and often nearly fills them with large gelatinous and varioufly undulated maffes, differing in fhape according to the rapidity or flownefs of the current. The color often varies in the fame mafs through every poffible fhade of a dull yellow, and Dr. Roth obferves that it frequently tinges ftagnant waters as if they were mixed with milk, and attributes this appearance to the tranfparency of the filaments, but as it is only obfervable on the furface of the maffes and where the filaments are much expofed to the fun, I fhould rather conceive it to arife from their having been bleached by its action.

C. ochracea is fo extremely fragile that the flighteft touch or even any confiderable agitation of the water breaks the filaments into a thoufand pieces, which are fo light as to remain fufpended in the water whilft the leaft agitation continnes, and then fubfide to the bottom in the form of an ochraceous powder. In this ftate only the plant can be examined, and prefents under the microfcope a multitude of fragments fo fmall that it is impoffible to afcertain the original length of the filaments, and fo extremely flender that under the higheft microfcopic power their thicknefs hardly appears equal to that of human hair of its natural fize. Two or more branches are frequently obfervable on the fame fragment. They are diffufe, moftly inflected, and difpofed without any apparent order. Diffepiments may occafionally be faintly diftinguifhed, but from the extreme tenuity of the filaments, not fo as to afcertain the length or nature of the joints, further than that they are perfectly cylindrical. No fructification has been difcovered.

In drying it adheres to both Glafs and Paper.

A. C. ochracea, natural fize.

B. Ditto, magnified 1.

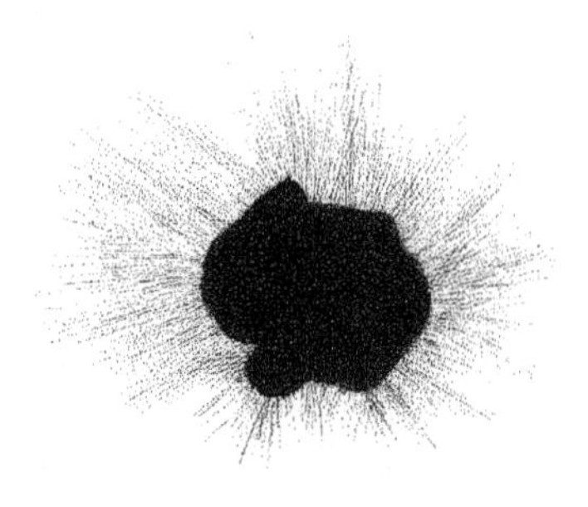

A

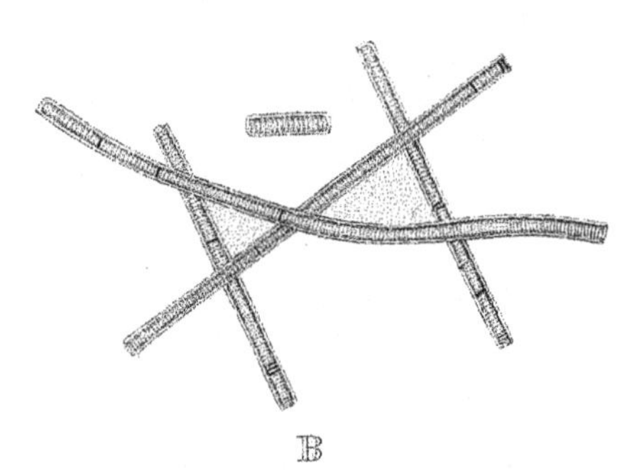

B

*Conferva fontinalis.*

J. Simpkins sc.

# CONFERVA FONTINALIS.

C. filamentis ſimplicibus cylindricis, truncatis, atro-virentibus, faſciatis; diſſepimentis obſcuris, articulis breviſſimis.

C. fontinalis. Sp. Plant. p. 1633. Fl. Ang. p. 592. Fl. Scot. p. 976. With. IV. p. 128. Fl. Dan. t. 651. f. 3. Roth, Fl. Germ. III. pars. 1. p. 593. Cat. Bot. II. p. 191.

Conferva minima biſſi facie. Dill. Muſc. p. 14. t. 12. f. 3.

Conferva fontalis fuſca omnium minima mollis. Ray Syn. p. 58.

In Rivers, Pools, Ditches, &c. common.

---

IN Dillenius's Herbarium the ſpecimen correſponding with t. 2. f. 3. is entirely deſtroyed by age, which perhaps renders it impoſſible poſitively to aſcertain the plant he intended, but his deſcription in the Hiſtoria Muſcorum ſo ſtrikingly correſponds with every appearance of the preſent plant, that I feel no heſitation in publiſhing it as that ſpecies. From ſpring to autumn it abounds in ciſterns, ditches, pools, rivers, and in ſhort in waters of almoſt every deſcription, generally floating in irregular maſſes on their ſurface. I gathered it lately in the King's Bath at Bath, where the temperature is 112 degrees, and it ſeemed not at all affected by the heat. In aërated waters, as Dillenius remarks, the ſurface of the maſs aſſumes an ochery color; in ditches and ſtagnant water it is frequently covered with decayed vegetable matter, in which it appears to delight, and I have found it in a rapid part of the River Lea, where its color was of a very dark and bluiſh green, and as it floated on the ſurface I at firſt miſtook it for C. diſtorta, for which, as Dr. Roth obſerves, it is very liable in this ſtate to be miſ-

9

taken, eſpecially by thoſe who have only ſeen the figure of that plant in the Flora Danica. It often may be found on pieces of decaying wood, &c. but I much doubt its at all adhering to them, as it does not appear to poſſeſs any root; it conſiſts merely of a filament equally obtuſe at both ends, and divided regularly by diſſepiments at very ſhort diſtances from each other.

Dillenius's C. gelatinoſa, omnium tenerrima, &c. * publiſhed in the ſecond Faſciculus of this work under the name of C. limoſa, I am inclined to ſuſpect is only the preſent ſpecies in a younger ſtate, and that when covered with water that plant in time riſes to the ſurface and aſſumes the appearances here deſcribed, the principal difference is in the ſize and color; in C. fontinalis the filaments are much larger, the color browner and not gloſſy as in C. limoſa; the joints alſo are far more diſtinct and more regularly diſpoſed. It is nearly allied to C. decorticans, but differs materially in ſize, in color, and in its much ſhorter joints, nor does it ever form the denſely matted patches, which give a ſtriking character to that ſpecies.

The growth of C. fontinalis is aſtoniſhingly rapid, and M. Adanſon's obſervations, from which I have given an extract in the deſcription of C. limoſa, apply equally to this and that plant.

When dried it alters its appearance but little, and adheres firmly to either Glaſs or Paper.

A. C. fontinalis, natural ſize.

B. Ditto, magnified 1.

* Hiſt. Muſc. p. 15. t. 2. f. 5.

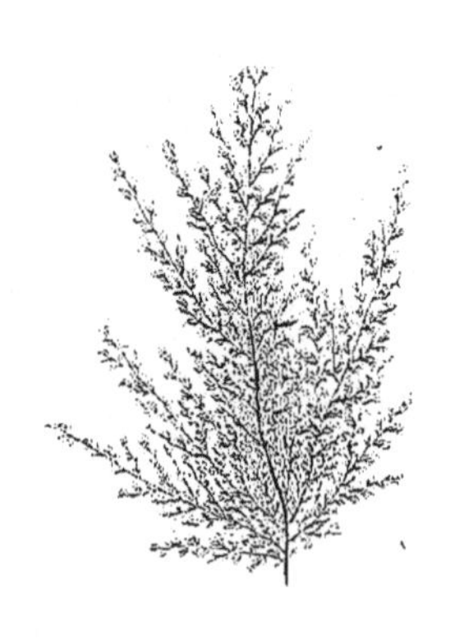

A

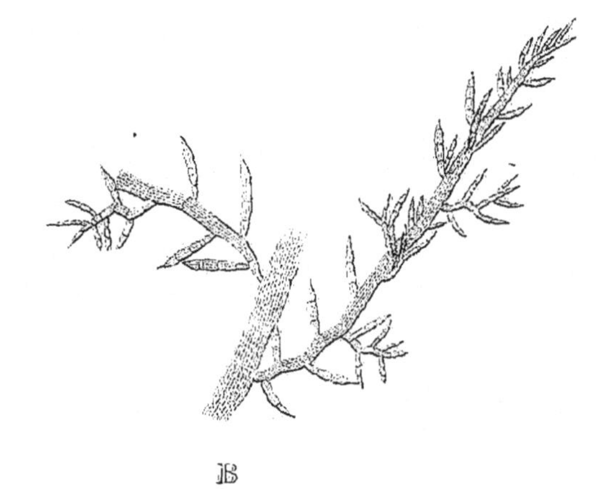

B

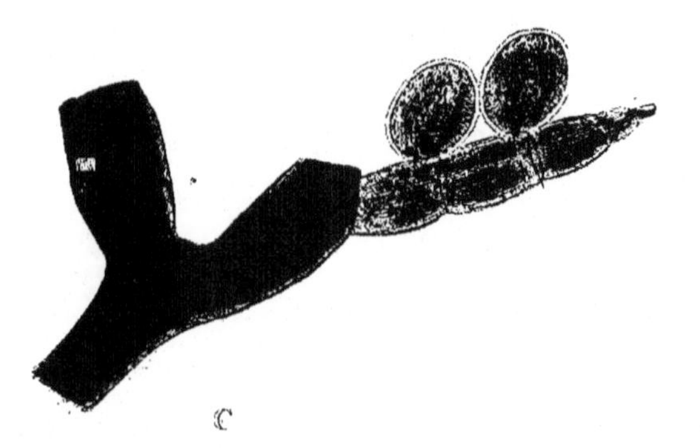

C

# CONFERVA TETRAGONA.

C. filamentis ramoſiſſimis ramulis faſciculatis brevibus ſimpliciuſculis; articulis ovato-cylindraceis, capſulis ſeſſilibus ſub globoſis.

C. tetragona. With. V. p. 405.

In the Sea, paraſitical on the Fuci. At the Bill of Portland, *Col. Velley.* In Caſwell and Llanglan Bays, near Swanſea. Shore at Weymouth, *D. Turner*

---

THIS elegant Conferva was diſcovered by Col. Velley and Mr. Stackhouſe at the Bill of Portland, and by them communicated to Dr. Withering, who firſt publiſhed a deſcription of it in an Appendix to the third edition of his arrangement of Britiſh plants.

C. tetragona is a ſpecies by no means found either generally or in abundance; it grows paraſitically on the larger Fuci in ſhrubby tufts, of a light purpliſh red color, ſeldom exceeding two inches in length. The root is a callus, common to many ſtraight and undivided ſtems, beſet with branches, not diſpoſed as in C. plumula on two oppoſite ſides only, but proceeding indiſcriminately from every part of it; neareſt the root ſhort, thence gradually increaſing in length to the center, and again decreaſing towards the ſummit, ſo that the general outline is irregularly ovate; they are again divided nearly in a ſimilar manner and are beſet with numerous cluſtered ſpine-like ramuli, extremely ſhort and for the moſt part ſimple, compoſed of joints ſomewhat reſembling thoſe of Fucus articulatus, the ultimate one terminated by an acute point. The fructification conſiſts of ſmall globular ſeſſile capſules arranged on the upper ſide of the armuli.

9

In drying it becomes darker, and adheres both to Glafs and Paper.

| | | | |
|---|---|---|---|
| A. | C. tetragona, | natural fize. | |
| B. | Ditto, | magnified | 3. |
| C. | Ditto, | ditto | 1. |

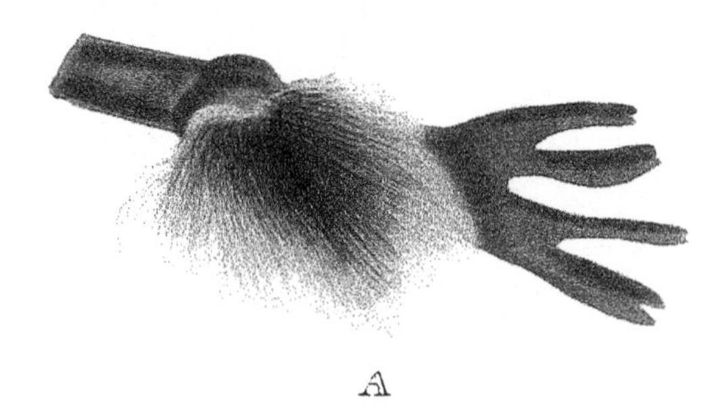

A

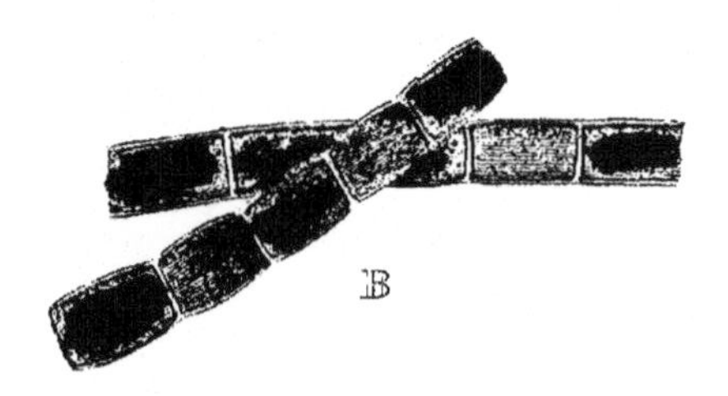

B

*Conferva fucicola.*

[illegible]

# CONFERVA FUCICOLA.

C. filamentis coefpitofis fimplicibus, obtufis; diffepimentis pellucidis parum contractis; articulis longiufculis.

C. fucicola. Velley's Marine Plants. pl. 4. With. IV. p. 136.

In the Sea; parafitical on Fucus nodofus & veficulofus, not uncommon.

---

MY friend Col. Velley firft difcovered the prefent fpecies, and gave a correct reprefentation and defcription of it among the colored figures of Marine plants with which he has favored the public. It is I believe far from uncommon on any of our fhores, generally growing on thick tufts on Fucus veficulofus, and fometimes, but much lefs frequently, on F. nodofus. Col. Velley jufty remarks that it does not feem to poffefs that indifference with refpect to places of growth which is ufual in Marine plants, as it has never been detected on rocks, fhells, or other extraneous bodies either by him or myfelf. An immenfe number of filaments generally grow together, thickly cluftered at the root, but while in the water diverging in a circular direction, and varying from four or fix lines to an inch in length: they are always unbranched and obtufe at the apices. The color is of a dirty yellow or brown, fomewhat gloffy when dried, and when viewed with a microfcope the whole filament exhibits a confiderable degree of tranfparency. The diffepiments are nearly colorlefs and flightly contracted; the joints are in length about equal to twice their thicknefs and are filled with minute granules, which may probably prove to be the fructification as no other has been difcovered.

In drying this plant adheres, though not very firmly, to either Glafs or Paper; its fubftance inclines to gelatinous.

A. C. fucicola, natural fize, growing on a piece of Fucus veficulofus.

B. Ditto, magnified 1.

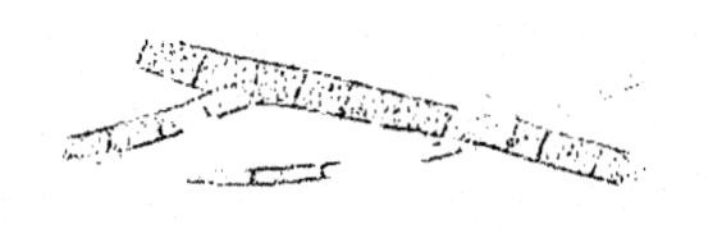

# COFERVA PROTENSA.

C. filamentis ramofiffimis, ramis diffufis, maximè elongatis, apicibus pellucidis articulis longiufculis.

In Rivulets and Springs growing on Stones, Wood, Reeds, and other aquatic vegetables; frequent about Swanfea.

---

THE prefent fpecies, though hitherto it has remained unnoticed, occurs in nearly every brook and rivulet about Swanfea, growing on ftones, fticks, graffes, reeds, and other aquatic plants: I have alfo met with it about Dover, and have no doubt it is by no means unfrequent in fuch fituations. The color is a light green; the filaments vary from two lines to half, and fometimes to three fourths of an inch in length, and are much branched. The branches are numerous, diffufe, and towards the apices fo lengthened out and pellucid that the termination of them is not eafily difcovered. The joints are of uncertain length, and are fhorteft in the ftem and longeft in the pellucid ends of the branches; with age they not unfrequently become inflated, and the jnices in drying often collapfe fo as to form two opake longitudinal lines parallel to each other, and leaving the remainder of the joints pellucid. The fructification has not been difcovered.

The Plant in drying adheres to either glafs or paper.

A. C. protenfa, natural fize.
B. Ditto, magnified 3.
C. Ditto, ditto 1.

9

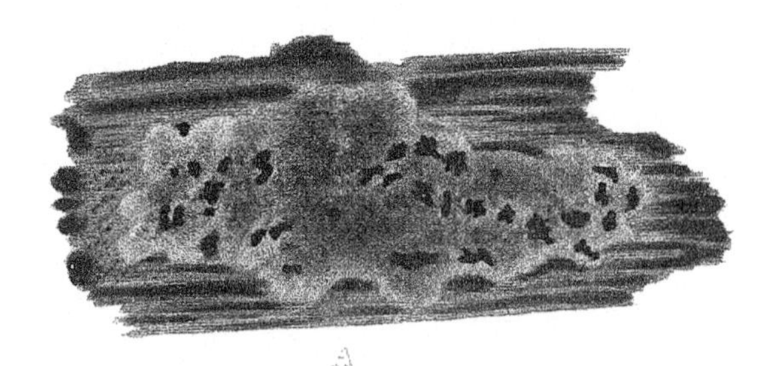
A.

B

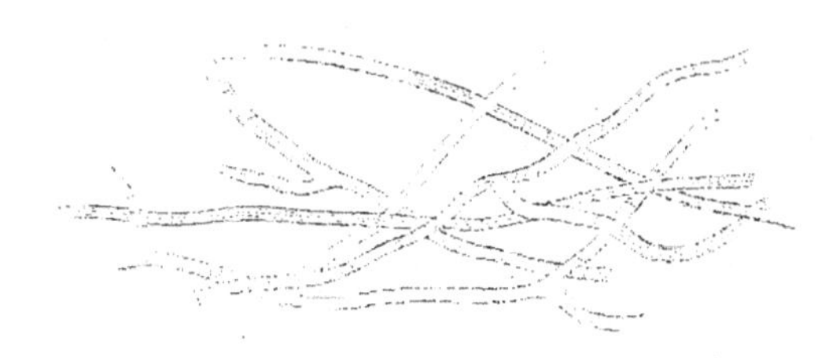

*Conferva ubiginosa?*

J. Simp. del.

# CONFERVA RUBIGINOSA.

C. filamentis ramofiffimis rigidis erectiufculis, ramis floruofis patentibus in maffam fub-folidam rubiginofam implexis-articulis longis.

On Rotten Wood.

---

FOR the prefent fpecies I have in vain fought through the Genus Byffus in moft of thofe authors who have defcribed that branch of Cryptogamia in which I conceive it moft probable that it would from its nature have been arranged. I therefore conclude that it has hitherto efcaped notice, and I have decided on giving it a place in this work from not being able to find any character which can diftinguifh it from the Confervæ. Indeed it appears to me, as far as my obfervations have hitherto gone, that the fame may be faid of all the Byffi filamentofæ.

C. rubiginofa grows on decayed wood in places where the light is nearly excluded, and forms irregular patches moftly about an eighth of an inch in thicknefs, and of a rufty brown color entirely deftitute of glofs. The primary filaments I have little doubt are repent, but fo mixed with the mould arifing from the decayed wood on which they grow, and afterwards fo denfely matted and entangled together that it is impoffible to feparate without tearing them, or to afcertain the nature of the ramification of the plant except towards the fummits. On examining a fection of the mafs it appears that from the creeping filaments rife upright ones which grow twifted together, and throwing out in every direction and without any regular order, patent flexuofe branches, every where of equal thicknefs, which are again entangled and matted fo as to form nearly a folid fubftance. Under the higheft powers of the microfcope, diffepiments are obſervable, which divide the filaments into joints in length about equal to four times their thickneſs. I have not been able to diſcover the fructification.

In drying it does not adhere firmly to either Glaſs or Paper.

| | | | |
|---|---|---|---|
| A. | C. rubiginoſa, | natural ſize. | |
| B. | Ditto, | magnified | 3. |
| C. | Ditto, | ditto | 1. |

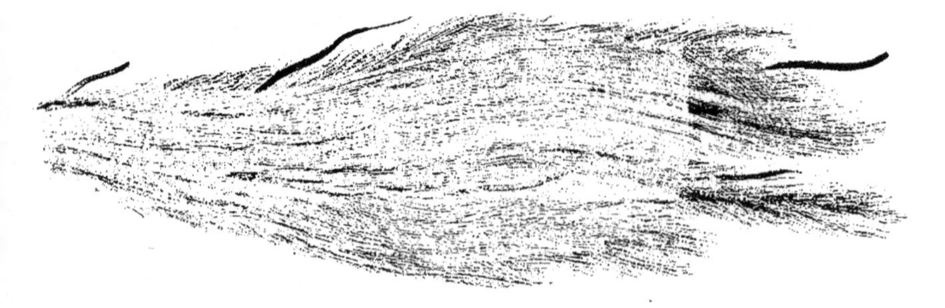

A

# CONFERVA DISSILIENS.

C. filamentis ſimplicibus ſtrictis fragilibus, diſſepimentis parum contractis plerumque ſolutis, articulis brevibus, in medio nigro-punctatis.

On Reeds and other aquatic vegetables in a Ditch on Cromlyn Bog, near Swanſea.

---

THIS ſpecies, which has not I believe been heretofore deſcribed, was firſt diſcovered by my friend and draftſman, W.W.Young, in the place above referred to, where it grows in great abundance on reeds and other aquatic plants. The manner of its growth is not ſo much entangled as in moſt of its congeners; its color is a dark green; the filaments are remarkably ſtraight and fragile; in length they are moſtly from three to ſix inches, and in thickneſs leſs than that of human hair. The diſſepiments are ſlightly contracted, and at theſe the filaments break, and the parts often remain connected at one extremity in the ſame manner as in C. pectinalis; the joints are in length about half equal to their thickneſs and on each ſide, both towards the diſſepiment and edge, are of a light green, whilſt the middle is of a darker color, ſometimes approaching to black, and this dark part at length becomes nearly round, and moſt pellucid at the center.

C. diſſiliens appears to be a link in the chain of ſubmerſed algæ, tending to connect C. pectinalis with C. nitida, rivularis, lucens and their congeners, from which it before ſeemed to be widely ſeparated. The preſent plant nearly approaches the nature and appearance of the latter in many reſpects, whilſt it claims an affinity with the former by its ſhort joints, and the manner in which the filaments break at the diſſepiments.

In drying it adheres very firmly to both Glaſs and Paper.

A. C. frangens, natural ſize.

B. Ditto, magnified 1.

*Conferva atro rubescens.*

# CONFERVA ATRO-RUBESCENS.

C. filamentis ramofis ftriatis, ramis elongatis fub-alternis, ramulis brevibus fubulatis fafciculatis; capfulis ovatis pedunculatis.

In the Sea, adhering to Rocks, Stones and Shells.

---

THE prefent fpecies appears to be far from uncommon on any of our fhores, and is occafionally found in large quantities in the bafins left by the tide. The length extends from four to fix or even nine inches; the color varies from a light purple to a dufky red, and becomes black with age, or by expofure to the air, as well as by drying; the root is a minute callus; the ftem folitary, of the thicknefs of fmall thread, repeatedly branched; the branches long, for the moft part alternate, and irregularly befet with awl-fhaped ramuli one or two lines in length, feveral of which are difpofed near each other fo as to give them a fafciculated appearance. The whole filament under the microfcope is ftriated in a beautiful manner by longitudinal veins, which arch over at or near the diffepiments, and at firft fight appear to form them; thefe veins are always in fome degree fpiral: the joints in the principal branches are in length frequently more than double their thicknefs, but in the ramuli the length and thicknefs are about equal; the capfules are ovate, and either lateral, on fhort fruit ftalks, or terminal at the end of the fmaller branches; other globular fubftances, imbedded in the joints, are alfo obfervable bearing a ftriking refemblance to the fuppofed fruit which conftitutes the variety β of Fucus coccineus, mentioned in the Synopfis of Britifh Fuci, and which, in my opinion, is occafioned by a collapfion of the juices.*

* I may take this opportunity to obferve, that I have found the globular capfules and this fuppofed fructification on the fame Frond of Fucus Coccineus.

10

This plant ſo thoroughly agrees with the deſcription of Hudſon's C. nigreſcens in the Flora Anglica, that it is with ſome heſitation I publiſh it under another name, but my friend Dawſon Turner informs me there are authentic ſpecimens of *that* plant extant, which prove the preſent to be an entirely diſtinct ſpecies.

The ſubſtance is ſtiff and rather rigid: in drying, the color becomes darker, and the plant adheres to paper, though but very ſlightly to glaſs.

A. C. rubro-ater, natural ſize.
B. Ditto magnified 2.
C. Ditto ditto 1.

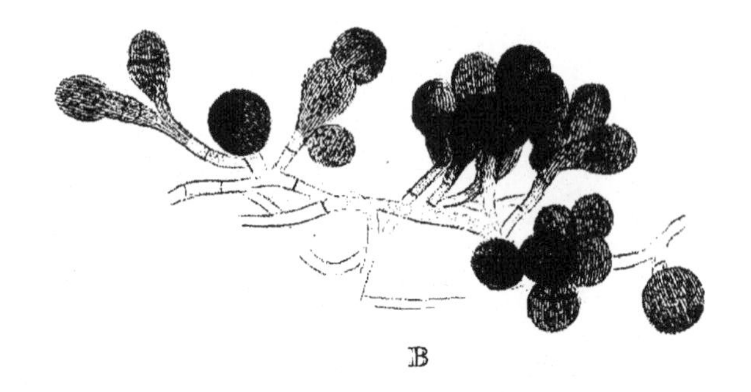


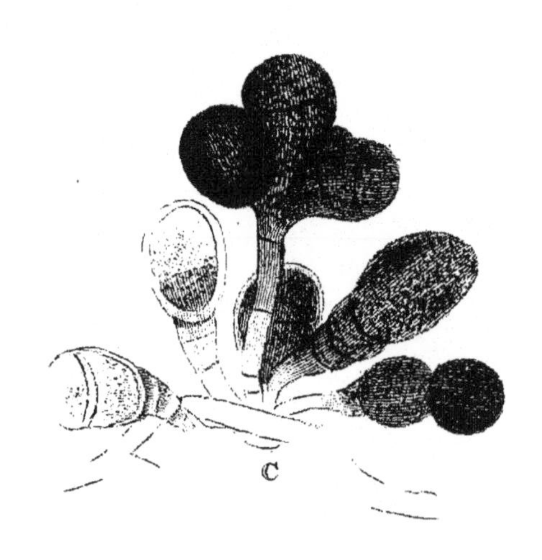


*Conferva multicapsularis.*

ing del!

# CONFERVA MULTICAPSULARIS.

C. filamentis minutis repentibus olivaceis, ramis erectis fimpliciufculis brevibus, apicem verfus incraffatis et capfuliferis; capfulis congeftis fphæricis.

On clayey banks in high and expofed fituations about Swanfea.

---

FOR the difcovery of this fingular Conferva I am indebted to my friend William Wefton Young, who found it growing on feveral parts of the Town-hill, near Swanfea: we have fince detected it together in other places in the neighbourhood. It grows on dry clayey banks, in expofed fituations, forming fmall irregular patches, which bear a confiderable refemblance to fome of the lichens. The color is a dark olive, often approaching to black, and forms a pleafing contraft with the light green of C. velutina, among which it is frequently found; the filaments are repent, thickly entangled, and very minute, fo that it is impoffible to afcertain their length; they throw out a number of fucker-like branches, from which numerous fhort upright branches arife, for the moft part fimple, but fometimes once or twice branched; thefe are thickeft towards their apices, and are thus frequently divided into two or more fhort palmated fegments, on each of which a capfule is placed. The joints are very long in the creeping ftems; they vary in the upright branches, being fhorteft at the bafe and longeft towards the fummit. When the juices from age have collapfed, or been dried up, the joints appear colorlefs, and filled with minute, ovate, pellucid granules, which I have alfo obferved in others of the fpecies that grow out of water. The capfules are difpofed at the end of the upright fhoots without any difcernible order: fometimes they are folitary, fometimes in clufters, and, not unfrequently, two or three may be feen apparently iffuing from each other; in moft of them a tranſverſe line is obſervable, at which the capſule divides when at maturity, and the ſeeds eſcape at the orifice.

In drying it adheres, though not very firmly, to either glaſs or paper.

A. C. multicapſularis, natural ſize.
B. Ditto magnified 2.
C. Ditto ditto 1.

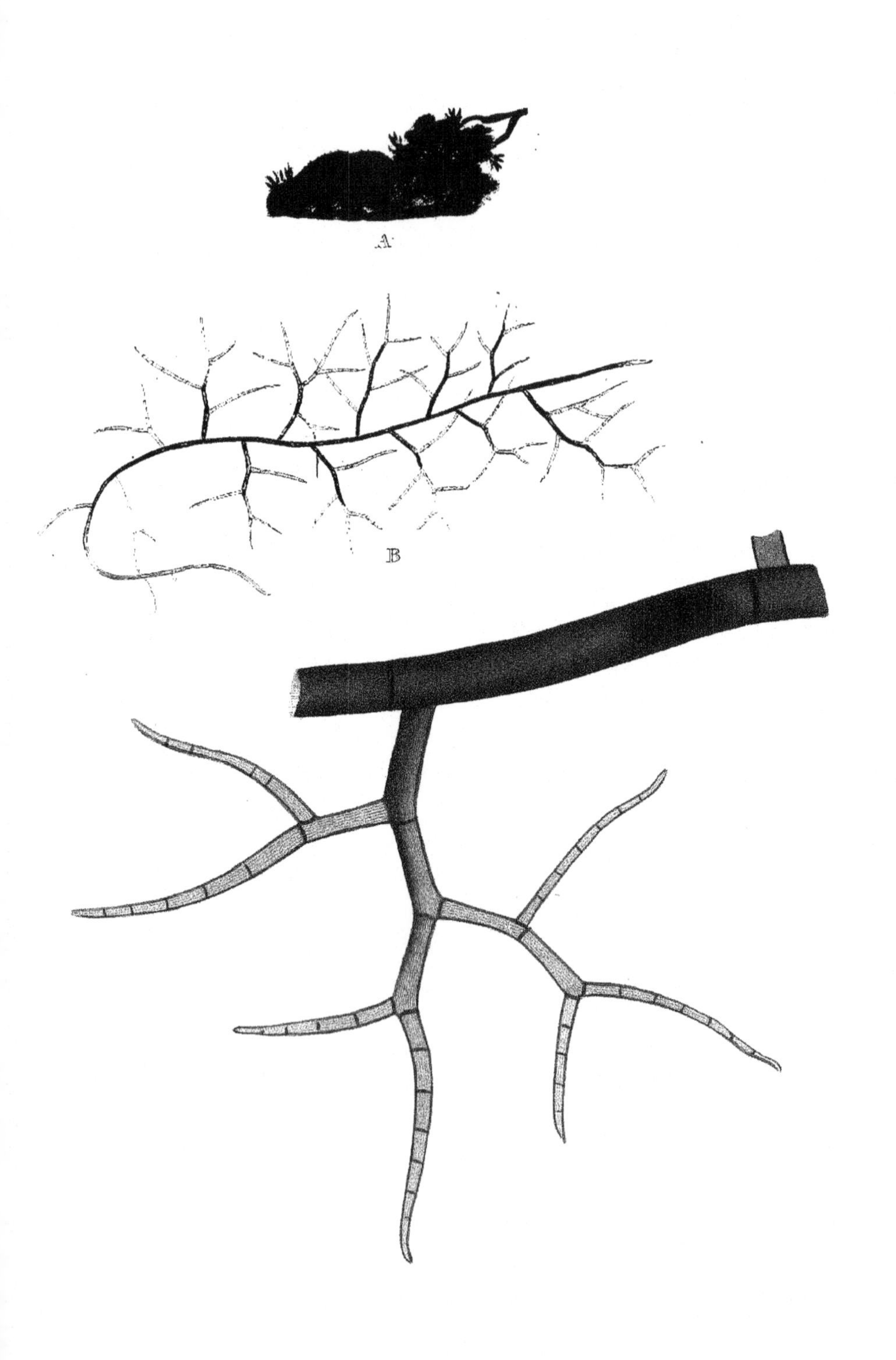
A
B

# CONFERVA CASTANEA.

C. filamentis repentibus ramofis fub-bipinnatis imbricatis implexis, pinnis pinnulifque alternis divaricatis, articulis longis.

On hedge banks in a lane on a high hill between the Gower and Lougher Roads, about four miles from Swanfea.

---

THIS fingular fpecies is found in great abundance on the fhady fide of a lane near Swanfea, and I am not aware of its having been elfewhere met with. It covers fticks, ftones, and earth, forming loofe patches of a brown-chefnut color. The ftem is creeping, and throws off feveral bipinnated decumbent branches, about a quarter or half an inch in length, which moftly grow over and become entangled with each other; the pinnæ and pinnulæ are regularly alternate, varioufly curved, and iffue at or nearly at right angles with the ftem and branches. The diffepiments are almoft black: in the principal branches the joints are very long, but they gradually become fhorter towards the ends of the ramuli No fructification has been difcovered.

In drying, the joints alternately collapfe, fo as to give the plant a fingularly beaded appearance; it adheres but flightly to either glafs or paper.

A. C. caftanea, natural fize.
B. Ditto magnified 3.
C. Ditto ditto 1.

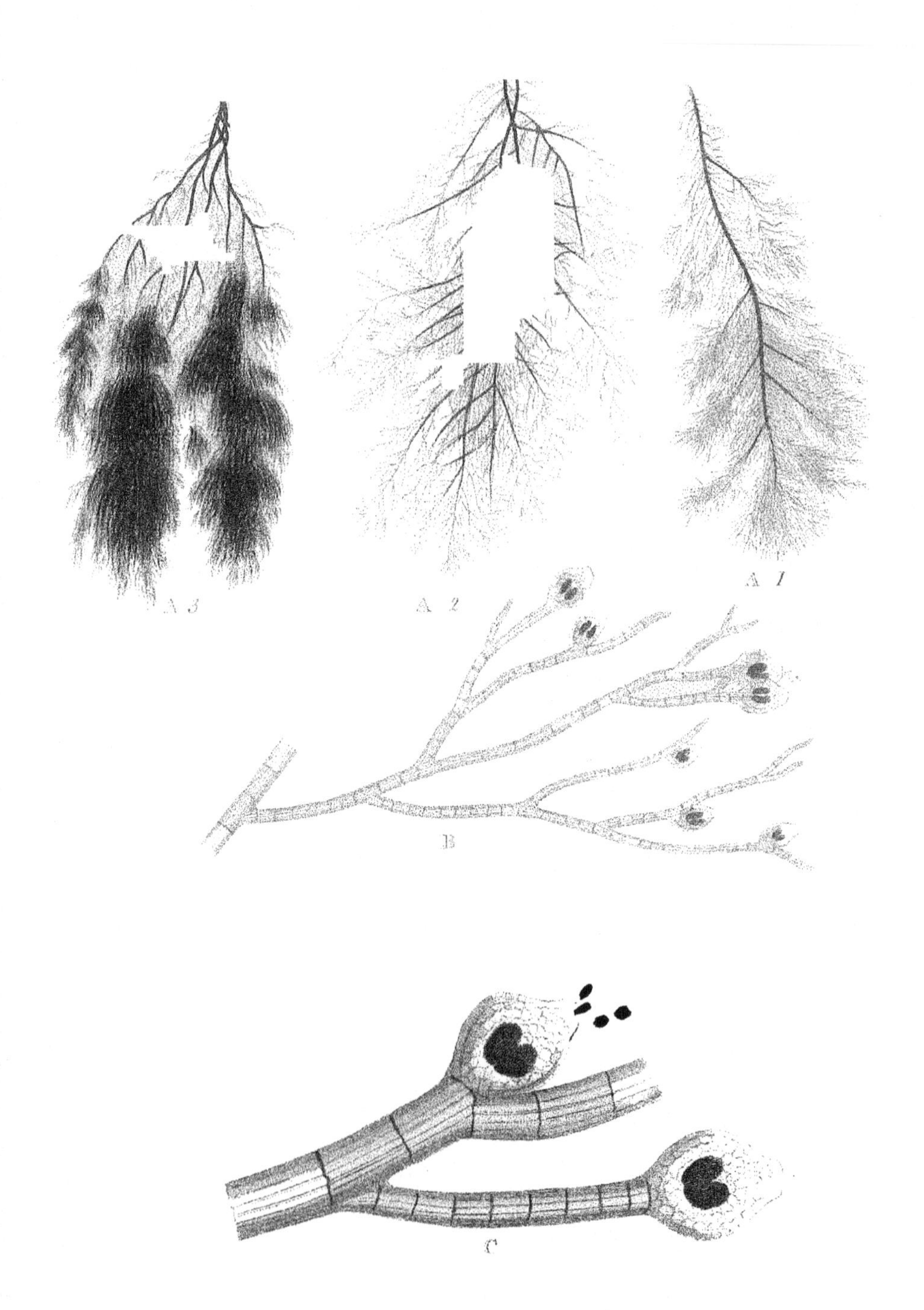


*Conferva [illegible]coides.*

# CONFERVA FUCOIDES.

C. filamentis ſub-cartilagineis ramoſiſſimis; ramulis dichotomis, diſſepimentis ex venarum anaſtomoſibus, articulis breviuſculis ſtriatis, capſulis ovatis ſub-ſeſſilibus.

C. fucoides. Fl. Ang. p. 603. With. IV. p. 141.

Ceramium violaceum. Roth. Cat. Bot. I. p. 150. III. p. 151. Fl. Germ. III. pars. 1. p. 462.

On Rocks and Stones in the Sea frequent.

---

AMONG the Confervæ few are ſo little known as the black marine ſpecies, which may be principally attributed to the ſhortneſs of Hudſon's deſcriptions, to his not having had any figures to which he could refer, and to the unfortunate deſtruction of his Herbarium. The difficulty in the preſent ſpecies has been removed by the kindneſs of my friends the Rev. Hugh Davies and Archibald Menzies, who, from among ſome authentic ſpecimens which they fortunately poſſeſs, have obligingly ſpared me two pieces marked 'C. fucoides' exactly correſponding with the plant here figured, as indeed does the deſcription in the Flora Anglica with ſome of the numerous appearances which it aſſumes in different ſituations and periods of its growth.

C. fucoides varies from two or three inches to a foot in length; its mode of growth is remarkably thick and buſhy; the color in the young plant is of a reddiſh-brown, becoming darker, and almoſt black with age; the root is a callus common to two or three irregular branched ſtems; the ſtem and main branches when the plant has arrived at maturity are in a conſiderable degree tough and horny; towards the ends they are repeatedly dichotomous; the diſſepiments, as

10

in C. byſſoides, appear to be formed by the arching over of the veins or nerves which are very obvious in the joints; the length of the joints varies; in the ſtem and principal branches it is three times their diameter, to which, in the ultimate ramuli, it is hardly more than equal; the capſules are ovate, either terminal or lateral; they are moſtly ſeſſile, but very ſhort fruit-ſtalks are ſometimes obſervable.

In drying it adheres but ſlightly to either glaſs or paper.

A. 1. 2. C. fucoides, natural ſize.
3. An old ſpecimen of the ſame, natural ſize.
B. C. fucoides, magnified 4.
C. Utimate Ramuli with Capſules, magnified 1.

A

B

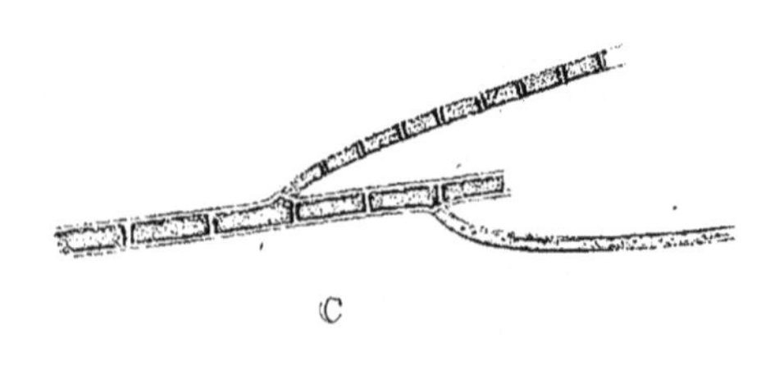

C

*Conferva rothii?*

# CONFERVA ROTHII.

C. filamentis erectis dichotomis brevibus densissimè cæspitosis phœniciis; ramis alternis, articulis breviusculis.

C. Rothii. Turton's System of Nature, VI. p. 1806.

C. violacea. Roth. Cat. Bot. I. p. 190, t. 4. f. 1. III. p. 224. Fl. Germ. III. pars. 1. p. 525.

On a Rock by the Sea-shore on the N. E. Coast of Anglesea, between Trosymarian and Penmain Park.—*Rev. Hugh Davies.*

---

I RECEIVED specimens of the present beautiful species from my friend the Rev. Hugh Davies, who first noticed it in Britain. He informs me, that it grows on a tophus, formed by the constant dripping of fresh water from an impending rock on the north-east side of Anglesea, between Trosymarian and Penmain Park, which is washed by the sea at spring-tides, and in rough weather. It appears to have been first discovered by Dr. Roth, on the piles placed on the shore, near Eckwarden, in the Duchy of Oldenburg. He described it under the name of C. violacea in his Catalecta Botanica, but as Hudson, in the Flora Anglica, had previously taken up a very different plant under that denomination, I have followed Dr. Turton, who, in his System of Nature, has altered its name to that of C. Rothii, in honor of its first discoverer.

C. Rothii grows in patches of various sizes, generally, according to Dr. Roth, affecting an oblong form. The color is a bright red, sometimes tending to brown, and changing, when dried, to a beautiful shining crimson; the filaments are very slender, frequently not more than three lines, and, I believe, never exceeding an inch in length; they are erect, densely matted together, and much branched; the

10

branches dichotomous, alternate, and moſt numerous towards the apices; the joints are cylindrical, and their length is about equal to twice their thickneſs; the interſtices pellucid. No fructification has been diſcovered. The Rev. Hugh Davies informs me that he has found this plant both in ſpring and autumn, but that the color is moſt brilliant in the latter ſeaſon.

C. Rothii has a conſiderable affinity to C. ſetacea and C. ſtricta; but to the naked eye its much ſmaller ſize, and, when magnified, the ſhortneſs of its joints will readily diſtinguiſh it from both theſe ſpecies

In drying it adheres to either glaſs or paper.

A. C. Rothii, natural ſize.
B. Ditto, magnified 3.
C. Ditto ditto 1.

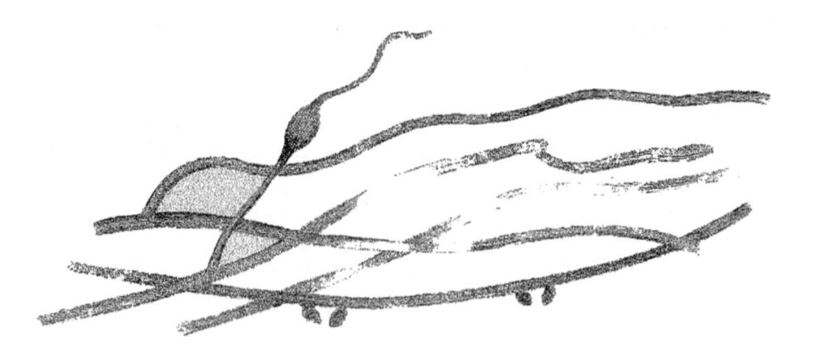

*Conferva vesicata?*

J. Sm

Published by L. W. Dillwyn, Sept. 1. 1806.

# CONFERVA VESICATA.

C. filamentis ramoſis ſub-articulatis, rigidis, veſiculis innatis ſolitariis ellipticis, filamento latioribus, capſulis ſubdidymis pyriformibus, breviter pedunculatis.

C. veſicata. Muller in Nov. Act. Pet. III.

C. burſata? Muller in Nov. Act. Pet. III.

In Fiſh Ponds at Knowle Park, and in a Stone Trough at the three mile ſtone on the Pensford Road, near Briſtol. *W. W. Young.*

---

MY Friend W. W. Young brought me the preſent intereſting ſpecies from the neighbourhood of Briſtol, and it ſo ſtrikingly agrees with Muller's figure and deſcription, as to leave no doubt of its being his C. veſicata. It grows in large buſhy maſſes at the bottom of the water. The filaments are ſo exceſſively brittle that it is almoſt impoſſible to aſcertain their length. They are cylindrical, every where ſtuffed with minute granules which iſſue from them when broken, and very rough to the touch; the branches are few, diſpoſed at a great diſtance from each other, and generally from an obtuſe angle with the ſtem. The ſtems and branches at irregular intervals are frequently ſwollen into bladder-like veſicles, four or five times broader than the filaments, and bearing a conſiderable reſemblance to thoſe of Fucus nodoſus. I obſerved one of their veſicles at the termination of a ſmall branch, as is repreſented in the figure, but, as Muller obſerves, I believe theſe very rarely occur. The diſſepiments appear very irregularly, though always at a great diſtance from each other, and towards them the joints are contracted at both ends; the capſules are pear-ſhaped, lateral, on ſhort footſtalks, and delicately reticulated with nerves; they are generally diſpoſed in pairs;

thefe capfules frequently occur on branches where no veficles are difcernible, and the plant then confiderably refembles C. burfata of Muller, which my friend D. Turner and myfelf found many years ago near Yarmouth, and which poffibly may not be a diftinct fpecies.

C. veficata agrees fo nearly in the nature of the filament, in its ramifications and joints with C. amphibia in an old ftate, that I think it rather doubtful whether future obfervations may not prove it to be only a variety of that plant. Its brittlenefs, and rigidity, and under the microfcope its fingular veficles will, however, readily diftinguifh it, and I have therefore thought it beft, and the moft certain way of avoiding future confufion, to follow Muller, and publifh it as a feparate fpecies.

In drying it adheres very flightly to either glafs or paper.

A. C. veficata, natural fize.
B. Ditto magnified 4.
C. Ditto ditto 1.

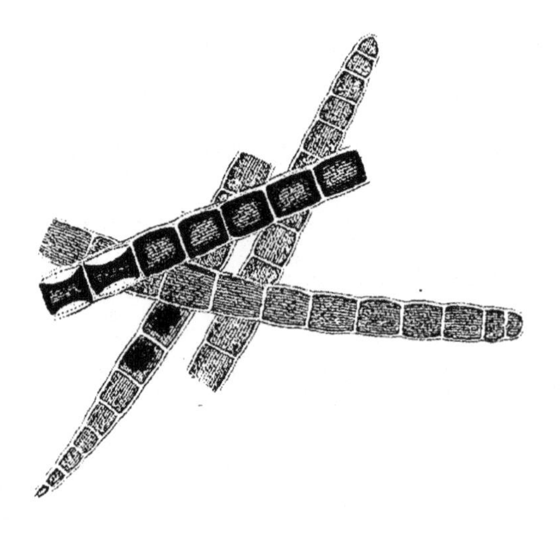

*Conferva curta.*

# CONFERVA CURTA.

C. filamentis cœſpitoſis ſimplicibus, ſub-cartilagineis, erectis brevibus utrinque attenuatis; diſſepimentis pellucidis parum contractis; articulis breviuſculis.

In the Sea, paraſitical on Fuci, not unfrequent at Swanſea.

---

THIS ſmall ſpecies, though it does not appear to have been heretofore noticed, is frequently to be met with on the ſhore at Swanſea, and I have reaſon to believe that it is far from rare in other parts of the kingdom. It grows paraſitically on Fuci, and forms roundiſh tufts ſo ſhort and ſtiff that they may be readily paſſed over as the remaining ſtumps of a paraſite, the greater part of whoſe filaments have been beaten off by the action of the waves. The color is a browniſh olive; the length, I believe, is ſeldom more than three or four lines. The filaments are ſimple; towards the root they are very ſlender, but become thicker as they approach the middle, and then again taper ſlightly towards their apices, which are rounded off and blunt. The diſſepiments are pellucid, and divide the filament into joints, whoſe length does not much exceed their thickneſs. No fructification has been diſcovered.

C. curta differs from C. fucicola, with which alone it can be at all confounded in the ſubſtance and color of the filaments: in the former they are rather of a horney nature and of an olive-brown color; the latter are remarkably flaccid and the color is more of a muddy yellow; the length and ſhape of the filaments are alſo materially different.

In drying it adheres to both Olaſs and Paper.

A. C. curta, natural ſize.
B. Ditto, magnified 3.
C. Ditto, ditto 1.

11

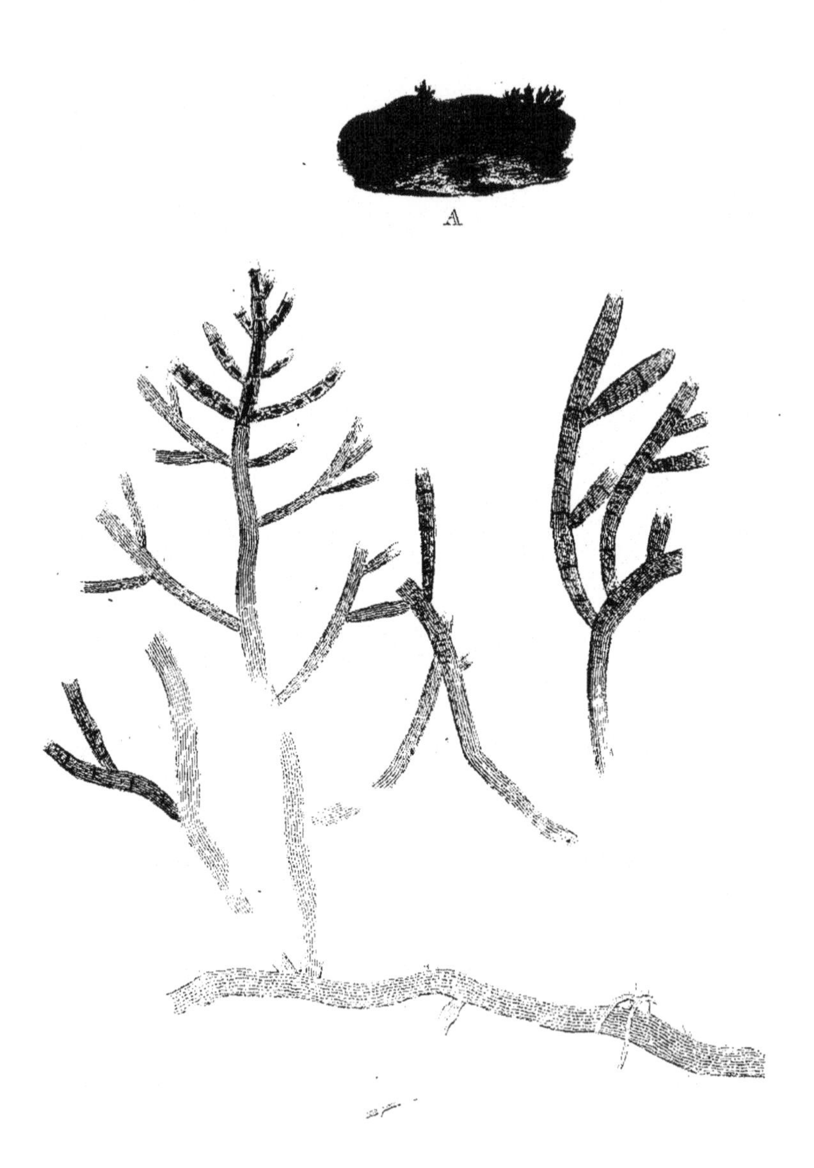


B

# CONFERVA VELUTINA.

C. filamentis repentibus ramofis implexis, ramis erectis fub-fecundis, curvatis, obtufis, articulis longis.

C. velutina. Eng. Bot. t. 1556.

C. varia. Roth. Cat. Bot. III. p. 301 ?

Byffus velutina. Linn. Sp. Pl. p. 1638. Fl. Ang. p. 605. Fl. Scot. p. 1001. With. IV. p. 144. Weis Crpy. p. 12. Roth. Fl. Germ. III. pars. 1. p. 562.

Byffus tenerrima viridis, velutum referens. Dill. Mufc. p. 7. t. 1. f. 14. Ray. Syn. p. 36.

Byffus terreftris viridis herbacea & molliffima, filamentis ramofis et non ramofis. Mich. Gen. p. 211. t. 89. f. 5.

On the ground in moift and fhady places.

---

C. VELUTINA grows moft frequently on moift fhady banks, and I believe is not uncommon in fnch fituations during the winter months and in the early part of fpring. It covers the ground with denfely matted patches, of a light or yellowifh green color, and frequently four or five inches in diameter. The filaments extend to a great length, throwing out roots below and branches from the upper fide; thefe branches are fhort, erect and matted together, fo as to bear a fancied refemblance to the pile of velvet from which the plant has derived its fpecific name. The branches are again twice or thrice divided with ramuli, for the moft part difpofed on the fame fide of the branch, but fometimes alternately; they are more or lefs curved and blunt at the apices. In the repent ftem and principal branches the diffepiments are hardly difcernable; the joints vary in length from twice to fix or eight times their thicknefs. Michæli's figure affords good reafon for believing that the fructification refembles that of *my** C. frigida figured at Plate 16, but I have not been able to difcover it.

C. umbrofa of Roth, figured at Plate 61 of this work, differs from C. velutina in its much darker color and more brittle nature. I however ftrongly fufpect that it is a variety only of this fpecies occafioned by its growth in a colder and boggy foil. The Conferva introduced by Dr. Roth in his Catalecta and Flora Germanica under this name, is an entirely different fpecies, and is the C. violacea of Hudfon and C. confragofa of the Flora Scotica.

In drying C. velutina adheres to both Glafs and Paper.

A. C. velutina, natural fize.

B. Ditto magnified 1.

* I may take this opportunity to obferve that the plant which I have figured under that name is not the C. frigida of Roth. Drs. Mohr and Weber, in their German edition of this work, firft corrected the error, and their correction is confirmed by the 3d vol. of the Catalecta Botanica lately publifhed. The fpecies which I erroneoufly figured under that name is there defcribed with the name of Ceramium Dillwynii.

# CONFERVA PALLIDA.

C. filamentis dichotomis, curvato-flexuofis, faftigiatis dichotomiarum angulis rotundis, articulis longiffimis.

On Yellow Ochre in Ifinglafs fize.

---

MY friend W. W. Young, having let fome yellow ochre remain about a fortnight in a pot of ifinglafs fize, found the furface of the ochre nearly covered by the prefent minute and interefting Conferva.

The color is of a light yellowifh brown: the filaments are confiderably finer than the fmalleft human hair, and are matted together into denfe leathery maffes, generally about an inch in length and of the thicknefs of a fhilling; they are much branched with repeated dichotomies of which the angles are uniformly rounded; the branches are fingularly flexuofe and curved all nearly of the fame length, and blunt and of a lighter color at the fummits: the length of the joints is irregular, in the ultimate branches they are equal to eight or ten times the diameter, and in the main branches are generally much longer. I have not been able to difcover any fructification.

It appears from the defcription in the Catalecta Botanica to be nearly allied to Roth's Conferva faftigiata, but in that fpecies the angle of the dichotomy is faid to be acute, and the joints very fhort and fomewhat beaded.

In drying it adheres to both Glafs and Paper.

A. C. pallida, natural fize.

B. Ditto, magnified 1.

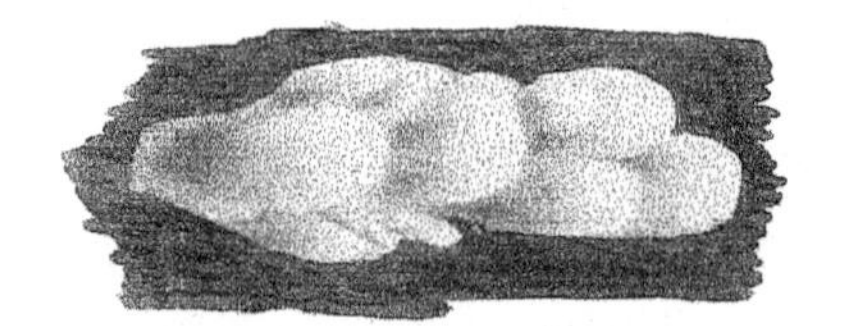

*Conferva lactea.*

# CONFERVA LACTEA.

C. filamentis ramosissimis, gelatinosis, lubricis; ramis virgatis alternis e quovis dissepimento; dissepimentis contractis; articulis longissimis, hyalinis.

C. lactea. Roth. Cat. Bot. I. p. 216. III. p. 292.

C. pusilla. Fl. Germ. III. pars. 1. p. 524.

In Ditches and Rivulets, growing on Stones, Wood, decaying vegetables, &c.

---

I HAVE found the present species in several places in Walthamstow and its neighbourhood, as also about Swansea, and I am inclined to think it is by no means unfrequent during the winter months. It grows on various substances at the bottom of ditches and rivulets, in gelatinous slippery masses, of a dirty white color, and varying from half an inch to three or four inches in length. The filaments are regularly branched at each dissepiment; the branches are alternate and so clustered as to give them a brush-like appearance. The dissepiments are of a dusky color, and divide the filaments into joints, whose length is various but never less than at least ten times their thickness, and they are slightly contracted and rounded at each end. Under the microscope they appear perfectly colorless, and this and their remarkable transparency will readily distinguish C. lactea from every other species with which I am acquainted. No fructification has been discovered.

In drying it adheres firmly to both Glass and Paper.

A. C. lactea, natural size.

B. Ditto, magnified 1.

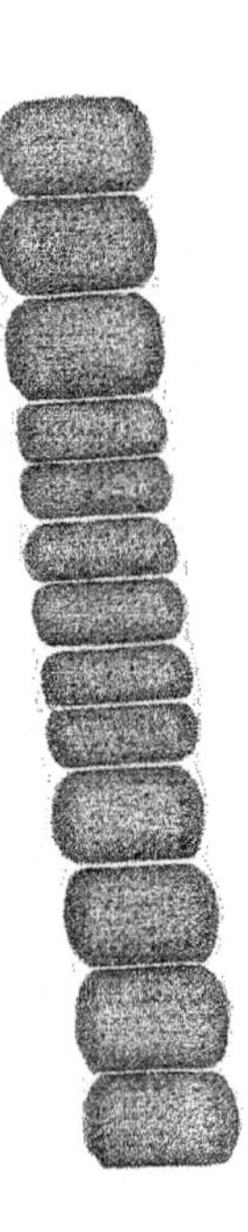

B

A

# CONFERVA ÆREA.

C. filamentis ſimplicibus rigidiuſculis ſtrictis; diſſepimentis hyalinis contractis, articulis oblongis brevibus.

On Stones in the Sea at Cromer, *D. Turner*. At the entrance of Laugharne Harbor; at Iſmael's Ferry, and other parts of the Carmarthenſhire Coaſt, *W. W. Young*. About Swanſea.

---

THIS ſpecies, which hitherto appears to have eſcaped the obſervation of any author, was, above four years ago, ſent by Dr. Goodenough to D. Turner, under the name of *C. Ærea*, and has ſince been found by my friend W. W. Young on ſeveral parts of the Coaſt of Carmarthenſhire; nor is it by any means unfrequent on the ſhore about Swanſea. Several filaments iſſue from the ſame root; they vary conſiderably in ſize. At the beginning of the winter before laſt I found one nearly of the thickneſs of a crow quill, but they are moſt generally about equal to large thread. They are invariably ſimple: their length is from ſix to fifteen inches; the color a dark or bluiſh green; they are brittle and rigid like C. capillaris, but not at all curled or entangled as in that ſpecies; the filaments contracted at the diſſepiments, which are remarkably pellucid and colorleſs; the length of the joints is leſs than their diameter, and two together often appear, whoſe united length is preciſely the ſame as that of one of the others, as if they had originally formed only a ſingle joint; they are rounded at each end, which gives the filament its beaded appearance. No fructification has been diſcovered.

When dried the filament aſſumes a more cylindrical form, and under the higher powers of the microſcope longitudinal fibres are obſervable. It adheres but ſlightly to either glaſs or paper.

A. C. concatenata, natural ſize.

B. Ditto magnified 2.

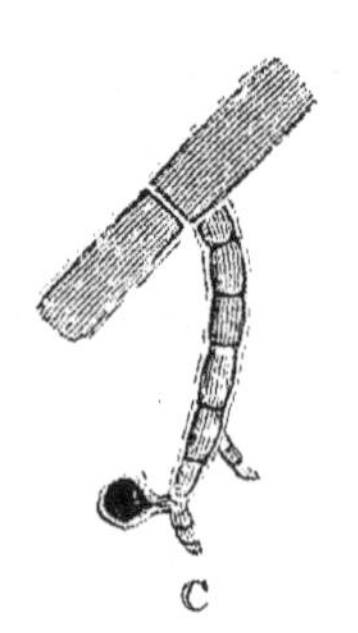


*Conferva setrica.*

# CONFERVA TETRICA.

C. filamentis decompofito-pinnatis, pinnis pinnulifque alternis, extremis curvatis; articulis longiufculis, capfulis fub-folitariis globofis pedunculatis.

On Fuci and on Rocks in the Sea. Common about the Mumbles and in other parts of the Peninfula of Gower.

C. TETRICA is extremely plentiful in the pools left by the tide on the coafts of the peninfula of Gower, where it grows either on the rocks or parafitically on the larger fuci. The root is a fmall callus from which feveral items arife, forming thick entangled bundles of a dull or brownifh red color, wholly devoid of glofs, and frequently attaining the length of fix or eight inches. The principal ftem in thicknefs is about equal to horfe hairs. The primary fhoots are difpofed without much obfervable order and of unequal lengths; they are winged with alternate branches, which are again pinnated with others alfo regularly alternate, and thefe are befet with fhort curved ramuli, of which the length is variable but always fhort in comparifon with that of the other branches. The joints are cylindrical; their length in the principal branches is at leaft equal to thrice their thicknefs, but it is much lefs in the fmaller ones. The capfules, of which feldom more than one occurs on any of the ramuli, are globofe, and placed on a fhort penduncle.

Although this plant fo ftrikingly differs in its greater fize and appearance in almoft every refpect from C. rofea, yet it is not eafy to find a fpecific difference when examined with the affiftance of a microfcope. The principal difference is then obfervable in the ultimate ramuli and in the difpofition of the capfules; the former in C. rofea are of regular lengths and truly pinnated; in C. tetrica they are again divided with fhort curved and fomewhat fpine-like ramuli. In C. rofea the capſules are numerous, ſeſſile, and regularly arranged on the upper ſide only of the pinnulæ, but in this plant they are on ſhort footſtalks, and ſeldom more than one is found on each ramulus, and that at or near its ſummit. The joints in the principal branches are alſo longer than C. roſea.

In drying C. tetrica does not adhere firmly to either Glaſs or Paper.

A. C. tetrica, natural ſize.
B. Ditto, magnified 3.
C. Ditto, ditto 1.

*Conferva setacea.*

# CONFERVA SETACEA.

C. filamentis sub-dichotomis, fafciculatis, ftrictis, virgatis, lubricis, ramis articulisque cylindraceis longiffimis; fructu laterali pedunculato.

C. fetacea. Fl. Ang. p. 599. With. IV. p. 137. E. Bot. XXIV. p. 1689.

C. marina gelatinofa, corallinæ inftar geniculata tenuior. Dill. mufc. p. 33. t. 6. f. 37. Turn. Tr. of Linn. Soc. VII. p. 107.

Corallina confervoides gelatinofa rubens, ramulis et geniculis peranguftis, R. Syn. p. 34.

On Rocks and Stones in the Sea, not unfrequent at the latter end of Summer and beginning of Autumn.

---

C. SETACEA has been obferved on moft if not all of our fhores, though in fome it is much more plentiful than on others. Where it inhabits it is almoft impoffible it fhould be overlooked, as its rich color muft attract the notice even of the moft incurious obferver. It conftantly grows in thick bundles, feldom exceeding four or five inches in length. The root is a fmall callus and gives rife to a number of rich crimfon filaments, generally more or lefs tinged with purple; they are branched with repeated dichotomies, the angles whereof are uniformly acute; the ultimate branches are long; the joints cylindrical; their length, efpecially in the main ftem, generally eight or ten times their breadth, and every where much longer than in any of its congeners. We are informed in Withering, on the authority of Col. Velley, that the fructification is in globular clufters on fhort lateral pedicles, but I* have never been fo fortunate as to meet with it. Col. Velley adds that it is rarely found.

The only two fpecies which can poffibly be confounded with C. fetacea are C. corallina and C. ftricta; from the former it differs in its more flender filaments and cylindrical joints; while its much lefs numerous branches, far longer joints, veinlefs filaments, fmaller fize, and brighter color, will readily diftinguifh it from the latter.

When this plant is placed in frefh water, a fcarlet liquor oozes from the joints; in drying it adheres to both glafs and paper. The colour is remarkably fugitive; it changes from expofure to the air to a dirty orange.

A. C. fetacea, natural fize.
B. Ditto, magnified 4.
C. Ditto, magnified 3.
D. Part of a fruit-bearing fpecimen, natural fize.
E. Portion magnified 2.
F. Seeds magnified 1.

* Since the above was written, Mr. W. J. Hooker has been fo kind as to favor me with a fketch of the fruit of this plant from a fpecimen in my friend D. Turner's collection, communicated to him by Mr. Templeton from the North of Ireland. There is fomething fo extraordinary and anomalous in the fructification, that I am unable to compare it with that of any other fubmerfed alga; the feeds are borne as Col. Velley defcribes them, but do not appear to be contained in a tubercle, and have a pellucid limbus more ftriking than in any fucus I am acquainted with. Mr. H. Davies has fuggefted that this plant, not C. rubra, as quoted by Hudfon, is the true C. flofculofa of Ellis.

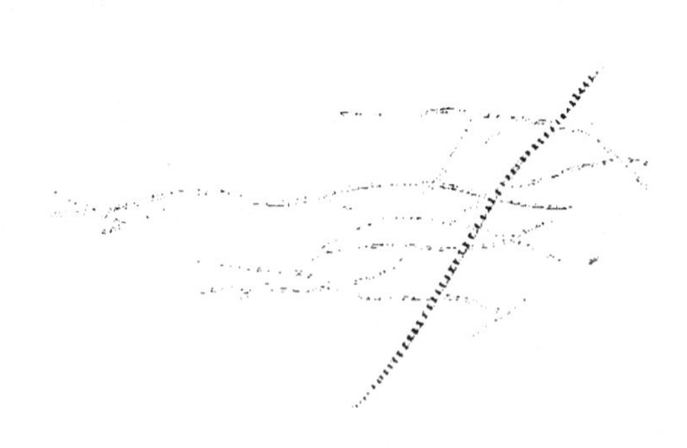

*Conferva typhloderma.*

# CONFERVA TYPHLODERMA.

C. filamentis ſub-ramoſis, denſiſſimè implexis, diſſepimentis obſcuris, articulis brevibus.

In Water which contained a Solution of Gum Dragon.

---

THE preſent Conferva was diſcovered by my friend William Weſton Young, in a bottle containing a ſolution of gum dragon in water, the ſurface of which it covered with a maſs of filaments ſo denſely interwoven as to form a cartilaginous film about two lines in thickneſs, and bearing a conſiderable reſemblance to the ſkin of a mole.

Their extreme tenuity and entangled growth makes it impoſſible to aſcertain the length of the filaments, which are generally ſimple, but a branch may be here and there obſerved—their color is a dull olive green. The diſſepiments are readily diſcernable, and are of a darker color than the reſt of the filaments, which they divide into joints, whoſe length is nearly but not quite equal to their thickneſs. No fructification has been diſcovered.

In drying it adheres firmly to both Glaſs and Paper.

A. C. typhloderma, natural ſize.

B. Ditto, magnified 1.

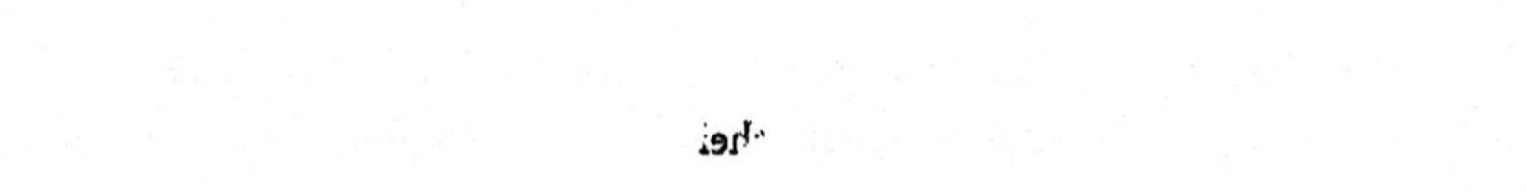

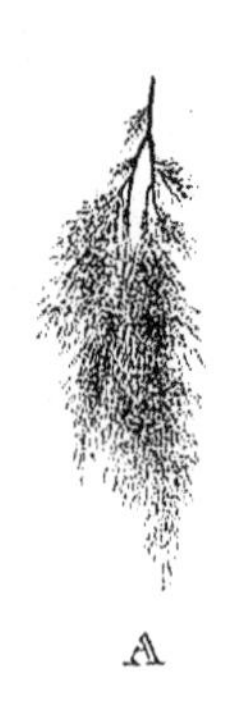


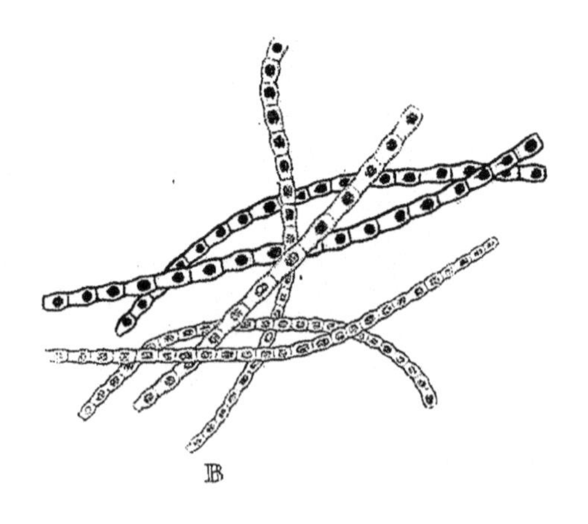


*Conferva carnea.*

# CONFERVA CARNEA.

C. filamentis ſimplicibus cæſpitoſis, ſub-nodoſis, carneis; articulis breviuſculis utrinque attenuatis; ſuccus in globulos ſolitarios congeſtus.

On Conferva in the River near Loughor, Glamorganſhire. *W. W. Young.*

---

IN September, 1805, Mr. Young brought me the preſent delicate ſpecies from the rocks in the Loughor river, where he gathered it, near to its confluence with the ſea. It grows on other confervæ, in looſe tufts, moſtly from a quarter to half an inch in length, and of a pale red or fleſh color. The filaments are ſimple, and taper in ſome degree both towards their root and apex, but terminate rather bluntly. The diſſepiments are of a dark color, and at regular diſtances from each other; the length of the joints in ſome filaments is about equal to twice their diameter; and in others the length and diameter are nearly equal. They are rounded off at both extremities, and moſt ſwollen towards the upper, ſo that when examined under the higher powers of the microſcope they bear ſome reſemblance to thoſe of Corallina officinalis. Among a number of young and apparently vigorous ſpecimens which Mr. Young examined, while they were quite freſh, he could not find one joint through which the juices were entirely diffuſed, and of which the greater part was not colorleſs, ſo as to induce him to believe that the red globules, of which one appears in each joint, are not the effect of a collapſion of the juices from age or expoſure to air, but natural to the plant in its moſt perfect ſtate; in ſome ſpecimens however which I examined when nearly freſh, I found that in the older filaments the red ſpot was conſiderably ſmaller in proportion to the ſize of the joint than in the younger ones, and I therefore preſume that they proceed entirely from a collapſion of the juices, which probably takes place in this more quickly than in moſt other ſpecies. I have not been able to diſcover the fructification.

There is no danger of its being confounded with any other ſpecies.

In drying it adheres to both glaſs and paper.

A. C. carnea, natural ſize.

B. Ditto, magnified 1.

*Conferva arbuscula.*

Published by L. W. Dillwyn, March 1st 1807.

# CONFERVA ARBUSCULA.

C. filamentis primariis incraffatis, inarticulatis, infernè denudatis, fupernè ramofiffimis; ramulis confertis, fubverticillatis, abbreviatis, ramofis, articulatis; articulis cylindraceis brevibus.

On fubmerfed calcareous Rocks near Ballycaftle, North of Ireland, *Mr. Brown*. Bantry Bay, *Mifs Hutchins*.

---

AMONG the various additions that have of late years been made to the lift of Britifh Confervæ, there is probably no fpecies more beautiful or interefting than the prefent, which was difcovered by Mr. Brown, fo long fince as 1800, in the habitat above mentioned. I find no traces of it in the works of any botanical writer upon the genus, nor have I ever met with any fpecimens befides thofe gathered by Mr. Brown, (to whom I am indebted for that here figured) except a fiugle one found by Mifs Hutchins, and preferved in the beautiful collection of my friend, Dr. Scott, of Dublin.

The root of C. arbufcula is, like that of moft other fpecies, a fmall callous difk, from which the filaments, as far as I have feen, arife in general fingly. Their height is about three or four inches. The leading fhoot, or ftem, (if I may ufe the expreffion) is as thick as packthread; nor, either in this, or the principal branches, have I been able to detect any traces of joints. It is naked and undivided near the root, at a fhort diftance from which it throws out branches, difpofed without any regular order, and much more clofely arranged in fome fpecimens than in others, the lower ones generally longeft, and the reft gradually fhorter, fo as to give the whole plant an irregularly ovate outline. Thefe branches are, like the ftem, naked near their bafe, and either fimple or again divided, clofely befet towards their apices with extremely fbort cluftered

12

ramuli, difpofed in a fubverticillate manner, irregularly branched, and very vifibly jointed, with cylindrical joints, of which the length is about equal to the diameter. The colour of this fpecies when frefh appears to be a beautifully deep-red brown; when dry it turns to a very dull brown, tinged with green, wholly devoid of glofs; and the plant at firft fight more refembles a battered fpecimen of C. fpongiofa infefted with fome minute parafite than any other Conferva. It adheres either to paper or glafs.

The fpecific name of this plant was given by Mr. Brown, and is excellently defcriptive of its mode of growth and general habit, which are not unlike that of many fpecimens of Hypnum alopecurum. T.

A. Conferva arbufcula, natural fize.
B. Summit of a branch, magnified 5.
C. Portion of ditto 4.
D. Ramulus 1.

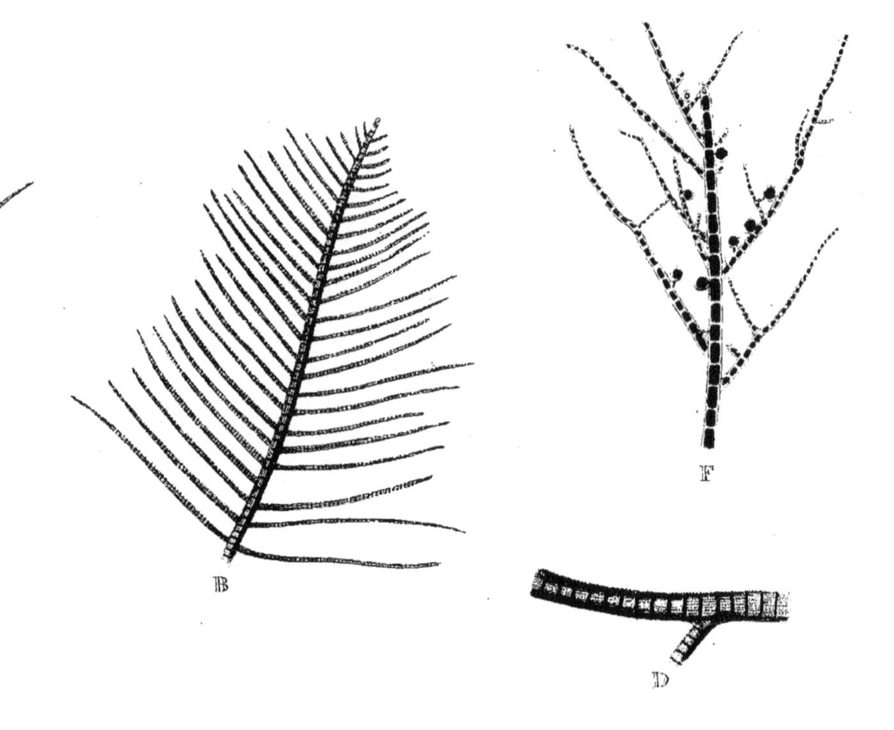


*Conferva pennata.*

J. Simpkins sculp.

# CONFERVA PENNATA.

C. filamentis ramofis; ramis pinnatis; pinnulis fub-oppofitis fub-horizontalibus, approximatis, ftrictis, diffepimentis obfcuris, articulis brevibus, tuberculis feffilibus fphæricis.

C. pennata. Fl. Ang. p. 604. With. IV. P. 142.

C. marina pennata. Ray, Syn. p. 59.

In the Sea, on Rocks, Fuci and Corallines. Common in Ireland, *Dr. Scott.* At Brighton, *Mr. Borrer.* Ifle of Wight, and Devonfhire and Cornifh Coaft, not unfrequent, *D. Turner.* In Anglefea, *Rev. Hugh Davies.* About Scarbro'. *Travis.* Near Forres in Elginfhire, *I. Brodie, Efq.* On the Mumble Rocks near Swanfea.

---

THE prefent fpecies, though far from uncommon, appears to have been remarkably ill underftood in general, and confounded by moft Botanifts with fmall varieties or broken pieces of Conferva fcoparia. Dr. Roth in the fecond Fafciculus of the Catalecta Botanica has referred it to his Ceranium pennatum, but in the third Fafciculus he corrects this error, and carries it properly to his C. cirrofa, with the defcription of which it does not however altogether accord.

C. pennata fometimes grows on rocks, but moft frequently on fuci or corallines, in bufhy tufts varying from half an inch to two inches in length; the color is olivaceous, becoming brown with age; the ftems are twice or thrice branched, but excepting the ultimate feries the branches can hardly be called pinnate; the pinnæ, which are long and fomewhat thorn-like, iffue almoft at right angles from the branches; their moft natural difpofition appears to be oppofite, and in fome plants two of them regularly iffue from each alternate joint, but in this refpect they are liable to great variation; the diffepiments are of a dark color, and divide the filaments into joints, whofe length does not exceed their thicknefs. The fructification which is drawn at F. from a dried fpecimen in the Herbarium of my friend D. Turner, confifts of globular feffile capfules on the branches.

For fome time I had confidered the plant figured at C. as a diftinct fpecies, and have diftributed a few fpecimens of it under the name of C. halecina. In this opinion I was joined by my much lamented friend the late Col. Velley, who had gathered it near Weymouth, but I have fince feen fome fpecimens in which the branches from the fame root have fo materially varied in the difpofition of their ramuli, as to convince me that it is a mere variety of the prefent fpecies. In this ftate it approaches fome of the varieties of C. littoralis, but may be readily diftinguifhed by its divaricated ramuli and more rigid nature. The fpecimen figured at B. was fent me by my highly refpected friend James Brodie, Efq. M. P. who gathered it near Forres in Elginfhire; where as alfo in other parts of the North, the pinnæ appear to be generally more regularly difpofed, than in the Southern parts of Britain. In Ireland the plant attains a larger fize than in England, as may be feen by the drawing at E. for which, and for that at F. I am indebted to the pencil of W. I. Hooker, Efq. of Norwich.

In drying it adheres, though not very firmly, to both glafs and paper.

A. C. pennata, natural fize.
B. Ditto, magnified 3.
C. Variety of ditto 3.
D. C. pennata, 1.
E. Irifh fpecimen, natural fize.
F. Specimen in fruit, magnified 2.

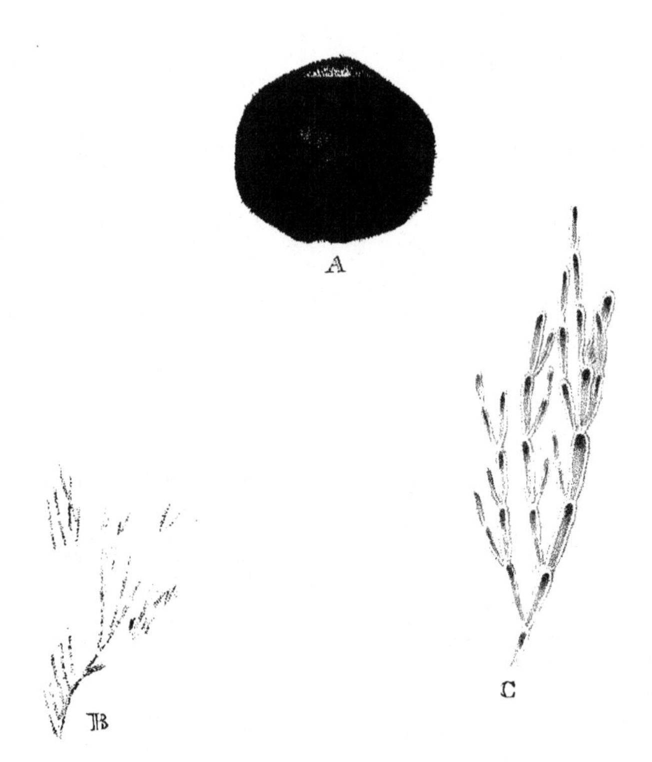


*Conferva ægagropila.*

# CONFERVA ÆGAGROPILA.

C. filamentis ramosissimis, e centro progredientibus, globum constituentibus; ramis ramulisque subsecundis strictis, obtusis; articulis longis, cylindraceis.

Linn. Sp. Pl. p. 1637. Fl. Suec. p. 436. Fl. Ang. p. 604. With. IV. p. 142. E. Bot. XX. p. 1377. Roth, Cat. Bot. I. p. 181. II. p. 212. III. p. 244. Roth, Fl. Germ. III. p. 517. Weber et Mohr Iter Suec. p. 71. t. 1. f. 7. *a. b.*

In Alpine lakes in North Wales, *Rev. H. Davies.* North of Scotland, *Mr. Brodie.* Prestwick Car, Northumberland, *Mr. Winch.* Culmere Pool, and Whitemere, Shropshire, *Rev. Mr. Williams.*

---

THE present species, and Conferva Arbuscula, I saw so little probability of being able to procure in a fresh state, that I have ventured, with respect to them, to deviate from my original intention, by giving representations of specimens that had previously been dried; for which I trust I shall not be blamed, as the British Catalogue can scarcely boast two more interesting individuals, or two without which a work on the Genus would be less complete.

Conferva Ægagropila is a native of mountainous lakes in different parts of Europe, having been found in Sweden, Norway, and both the North and South of Germany. Dr. Roth has enumerated three varieties, of which I am not aware that more than one has hitherto been met with in England. Its size is uncertain, varying from that of a pea to a large walnut. The filaments always originate from a center, and extend with repeated ramifications to the extremities, preserving an equal height, so that the form of the whole plant is constantly globular, in which, as far as my knowledge extends, no other species of Conferva refembles it. No root, however, has yet been detected, nor any folid body within the mafs, to which the filaments might originally have been attached. The mode of ramification feems fomewhat uncertain, but the branches and ramuli are principally difpofed on one fide; they are always ftraight, and their apices are regularly obtufe. The length of the joints is about equal to three times their diameter; in a recent ftate they are perfectly cylindrical; but, when dried, the green matter collapfes as in moft others of this tribe, and never afterwards recovers itfelf by immerfion. The colour of this plant is a dark, but pleafant, green, deftitute of glofs. In drying it does not in the leaft adhere to either glafs or paper. It is fufficiently known that it derives its fpecific name from its refemblance to the hairy balls found in the ftomachs of goats. For the fpecimen here figured I am indebted to my excellent friend, Mr. Brodie.

Many Botanifts have been led into error refpecting C. Ægagropila, from the circumftance of fragments of C. capillaris being occafionally found rolled up by the tide fo as greatly to refemble that fpecies at firft fight, though it can fcarcely be neceffary to fay that the difference may immediately be detected on looking more clofely at them. Of thefe I have feen vaft numbers at different times on the fhores of the river at Yarmouth, but they are by no means of frequent occurrence. Is it poffible that C. Ægagropila itfelf fhould derive its globular form from a fimilar circumftance? *T.*

A. C. ægagropila, natural fize.
B. A branch, magnified 4.
C. A portion of ditto 2.

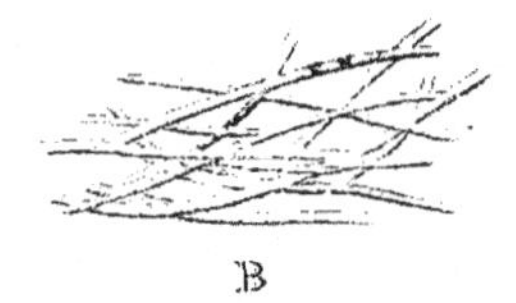

B

*Conferva phosphorea.*

Simpkins sculp

Published by L. W. Dillwyn June 1st 1807.

# CONFERVA PHOSPHOREA.

C. filamentis breviffimis ramofis, adfcendentibus, denfiffimè in cruftam uniformam implexis, violaceis; diffepimentis obfoletis, articulis longiufculis.

Byffus phofphorea. Sp. Pl. p. 1638. Fl. Ang. p. 605. Fl. Scot. p. 1000. With. IV. p. 143. Fl. Germ.. 564.

Auricularia phofphorea. Sowerby's Englifh Fungi. III. f. 350.

Byffus lanuginofa violacea liguis adnefcens. Ray. Syn. p. 56. Dill. Mufc. p. 54. t. 1. f. 6.

Byffus purpurea delicatiffima, arborum corticibus adnafcens, breviffimis & tenuiffimis filamentis. Mich. Gen. p. 211. t. 90. f. 3.

Byffus cœrulea cæfpitofa crifpa. Hall. Hift. p. 2102.

Fungus violaceus herpatis modo liguis irrepens. Ray. Hift. III. p. 23.

On decaying wood.

---

THE Byffi filamentofæ; moft of which I have had an opportunity of examining, fo nearly refemble each other in ftructure and mode of growth, that the fame reafons for which aurea and purpurea were transferred to the Confervæ, will equally apply to them all, though probably to none lefs than to the prefent fpecies, which it is not without confiderable reluctance that I admit among the Confervæ, regarding it as a plant with the true hiftory of which we are at prefent very little acquainted. Mr. Sowerby has claffed it among the Auriculariæ, and obferves that in its moft perfect ftate it feems to belong to that tribe, in which opinion he may poffibly be correct, as I have myfelf feen fpecimens of which the edges were of a pale ftraw color like many of thofe Fungi, and inclined to

curl off the wood they grow upon. At the ſame time as no author has noticed it in the ſtate in which it is here repreſented, I feel a pleaſure in probably contributing to throw ſome light upon it, and I leave it to future naturaliſts to determine its place in the ſyſtem.

I received the ſpecimen from which the preſent drawing was made from my friend T. W. Dyer, Eſq. who gathered it in Somerſetſhire. It grows on decaying wood, in patches of various ſizes, and of a beautiful and vivid violet color, which is permanent many years after it has been dried. The filaments are ſo extremely ſhort and much interwoven that the patches to the naked eye greatly reſemble the cruſt of a lichen, but their filamentous nature is in moſt ſpecimens* obſervable with the aſſiſtance of a common glaſs. The minuteneſs is ſuch that it is impoſſible to ſeparate them, ſo as to aſcertain the preciſe length or the frequency of their ramifications, but I apprehend the former rarely exceeds half a line, and that there are ſeldom more than one or two branches on each filament. The diſſepiments are by no means ſo eaſily diſcerned, or ſo regular as in C. purpurea, but are here and there obſervable, and divide the filaments into joints, of which the length exceeds the diameter. No fructification has been diſcovered.

A. C. phoſphorea, on decaying wood, natural ſize.
B. C. magnified 1.

* I have examined ſome ſpecimens in which I could not detect them at all, and I therefore feel ſome doubt whether they may not be peculiar to a certain age or ſtate of the plant.

*Conferva orthotrichi.*

# CONFERVA ORTHOTRICHI.

C. filamentis cœſpitoſis, pulvinatis, rigidiuſculis, fragilibus, ramoſis; ramis ſub-alternis, obtuſis; articulis brevibus, diametrum vix ſuperantibus.

C. muſcicola. E. B. XXIII. t. 1638.

On trees in the New Forest, Hampſhire, growing on *Orthotrichum ſtriatum*. C. *Lyell*, Eſq.

---

THE name of C. *muſcicola*, given to this ſpecies in Engliſh Botany, having been previouſly beſtowed upon a very different plant in Dr. Weber and Dr. Mohr's admirable Swediſh Tour,* and ſubſequently in Dr. Roth's Catalecta Botanica, I have been under the neceſſity of adopting a new one, and have with the concurrence of Dr. Smith, taken that of C. *Orthotrichi*, as the plant has at preſent been found upon no other tribe of moſſes. For the ſpecimen here figured I am indebted to Mr. Sowerby, to whom it was ſent by Mr. Lyell, the only perſon who appears to have yet found it in England, except indeed, as I ſuſpect, the curled appearance of *Orthotrichum ſtriatum*, mentioned in the Muſcologia Hibernica as the variety β, ſhould prove to be the beginning of it.

C. *Orthotrichi* grows in very thick entangled tufts on the upper branches of moſſes, having its roots in the leaves and ſtem, which it often ſo completely covers as to leave ſcarcely any part of them viſible. It is of a rich cheſnut color, dull and without gloſs when dry. The filaments are not above two or three lines high, erect, repeatedly branched; the branches generally diſpoſed at ſome

* *Reiſe durch Schweden*, p. 60. t. 1. f. 3. The Conferva here figured ſo nearly reſembles the *C. caſtanea* of this work, that I am apprehenſive they are not diſtinct, and I am ſorry I was unacquainted with Dr. Mohr's plant when I publiſhed my own Before, however, I conſider them as certainly the ſame, I ſhall hope for ſpecimens from that able botaniſt.

13.

diſtance from each other, in an irregularly alternate manner, ſhort, blunt, ſimple, iſſuing from the ſtem at obtuſe angles, and pointing upwards. The joints throughout the whole plant are uniform, their length ſcarcely greater than their breadth, and with ſomewhat of a beaded appearance. Theſe circumſtances will be ſufficient always to diſtinguiſh it from *C. caſtanea*, t. 72, to which at firſt ſight it bears a ſtrong reſemblance. Great care is neceſſary not to confound either of theſe plants with the radicles, which ſhoot out of the ſtems of moſt ſpecies of moſſes that grow in moiſt places, and are particularly abundant on *Bartramia fontana* & *Bryum paluſtre*. C. *Orthotrichi* is alſo very nearly allied to *C. Acharii* & C. *rubicunda* of Roth, the latter of which may probably be the C. *ilicicola* of Engliſh Botany.

The texture of C. *Orthotrichi* is rigid and brittle; and in drying it adheres neither to glaſs nor paper. *T.*

A. A ſtem of *Orthotrichum ſtriatum* nearly covered with *C. Orthotrichi*, natural ſize.

B. Summit of ditto, magnified 6.

C. Leaf ditto 5.

D. Ditto ditto 4.

E. C. *Orthotrichi*, ſeparate 1.

A
B

# CONFERVA PELLUCIDA.

C. filamentis erectis, ſtrictis, ramoſiſſimis; ramis plerumque ternis, obtuſis; articulis cylindraceis diametro quintuplo longioribus.

*C. pellucida.* Fl. Ang. p. 601. With. IV. p. 139. E. B. XXIV. f. 1716.

C. prolifera β. tenuior. Roth, Cat. Bot. III. p. 247.

On rocks, and ſtones in the ſea in Devonſhire, Cornwall, Suſſex, and Hampſhire. *Hudſon.* On the beach at Yarmouth.

---

THIS Conferva, though ſaid by Hudſon to be a native of ſo many counties, does not ſeem by any means a common ſpecies, and is certainly one of thoſe which are leaſt underſtood by modern botaniſts. How far Dr. Roth is right in referring it as a variety to his *Conferva prolifera* is a point I can by no means attempt to decide, as that plant is not a native of the Britiſh ſhores, and every perſon acquainted with this tribe muſt be aware how impoſſible it is to ſpeak with confidence from dried ſpecimens. I rather incline, however, from their different habits to think he is miſtaken.

The root of C. *pellucida* is a ſmall diſk, from which the filaments riſe in general ſingle; ſimple and naked at their baſe, but ſoon becoming branched, and afterwards ſo repeatedly divided, that the appearance of the plant towards the apices is remarkably buſhy. Their length is ſix or eight inches; their texture ſtiff, wiry, and elaſtic when freſh, but ſoon turning flaccid; their color a remarkably pleaſing, pale, ſubdiaphanous green, which is permanent even after drying. The branches are chiefly ternate, though ſometimes oppoſite, or even alternate; very ſtraight; between erect and patent; the apices bluntiſh. The length of the joints is about five or ſix times greater than their breadth; they are either quite cylindrical, or very ſlightly incraſſated upwards: the diſſepiments are dark and narrow in a recent ſtate, but, as the plant decays, grow pellucid, from the collapſing of the juices.

For the drawing of this plant, as well as of the following, C. *Orthotrichi*, I am indebted to my friend, W. J. Hooker, Eſq.

In drying it does not adhere at all to glaſs, and very ſlightly to paper. *T.*

A. Conferva pellucida, natural ſize.
B. A ſmall branch, magnified 5.
C. A part of ditto 3.

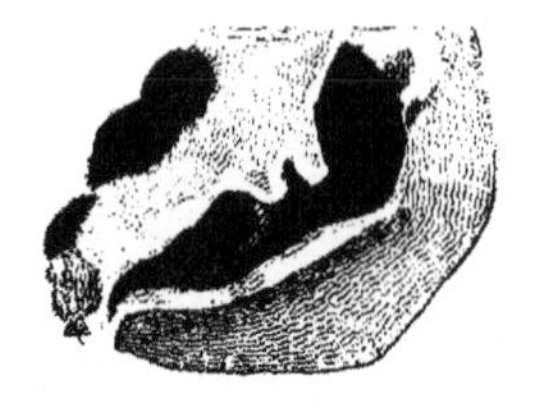
A

B

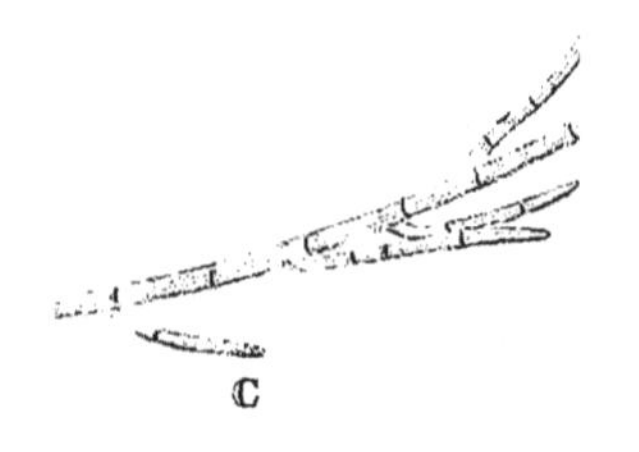
C

*Conferva chalybea.*

# CONFERVA CHALYBEA.

C. filamentis pulvinatis, ramofis, tenuiffimis, ftrictis, erectis, faftigiatis; ramis fub-alternatim fecundis, adfcentibus, obtufis; articulis cylindraceis, longis.

C. chalybea. Roth. Cat. Bot. III. p. 286. Tab. 8. f. 2.

On Flints in Winterbourne Stream at Lewes, Suffex. *Wm. Borrer, jun.* Efq.

---

PROFESSOR Mertens firft discovered the prefent delicate fpecies in the neighbourhood of Bremen, and communicated it to Dr. Roth, who has publifhed it with a good figure in the third volume of his Catalecta Botanica.—Mr. Borrer has fince added it to the Britifh Flora, having found it in Winterbourne Stream at Lewes, Suffex, and to him and Mr. Turner I am indebted for the fpecimen here figured. It grows on *flint-ftones* in little tufts about a quarter of an inch in length, and of a blackifh-green color, gloffy when dry. The filaments, which are repeatedly branched, are erect, ftraight, of equal height, and very flaccid and flender throughout. The branches are placed at uncertain, generally confiderable, diftances from each other; and iffue from the ftem fo as to form an obtuse angle, but immediately curve inwards, and then rife in a more or lefs upright direction; their difpofition is far from regular, but they are frequently difpofed on oppofite fides in alternate parcels of two or three. The ramuli are always placed nearer to each other than the main branches, and I have frequently obferved more than one proceeding from the top of the fame joint; they are blunt at their apices; the diffepiments are readily obfervable with a microfcope, and divide the filaments into perfectly cylindrical joints, of which the length is generally from four to fix times greater than the diameter.

C. chalybea is moft nearly allied to C. vivipara, but the defcription and figure here given will fufficiently prove it diftinct.

In drying it adheres to both glafs and paper, and more readily revives when immerfed in water than moft other fpecies.

A. C. chalybea, growing on a flint, natural fize.
B. ditto magnified 3.
C. ditto ditto 1.

# CONFERVA FUSCO-PURPUREA.

C. filamentis ſimplicibus, tenuiſſimis, rectis, ſub-faſciculatis, ætate inæqualiter toroſis; articulis brevibus utrinque ſub-pellucidis, demum ſerie globulorum cinctis.

On lime-ſtone rocks in the ſea about high water mark in the neighbourhood of Dunraven Caſtle. *W. W. Young.*

---

FOR the diſcovery of this Conferva I am indebted to my friend, W. W. Young, who found it growing with C. Rothii and another ſpecies not yet deſcribed, on the lime-ſtone rocks, a little below high water mark, in the neighbourhood of Dunraven. It may be worth remarking that theſe rocks which produce ſeveral Confervæ, that we have not been able to find on the Mumbles or other lime-ſtone in the weſtern parts of the County, are of a different ſort of lime-ſtone, and of that kind which I am informed is uſually called *lias ſtone*, and are ſimilar in quality to thoſe of which large quantities are exported from Aberthoir, and uſed for the ſame purpoſes as Dutch terrace. This and many other obſervations which I have made ſtrongly tend to confirm the opinion of my friend Dawſon Turner,* that the roots of the marine Algæ are not merely intended by nature to fix them to their places of growth, but that they are alſo uſeful as organs of nutrition, although the hardneſs of the ſubſtances on which many of theſe plants grow has led many botaniſts to ſuppoſe the contrary.

Mr. Young informs me that C. fuſco-purpurea frequently grows in very large patches, ſo as to cover the rocks for two or three ſquare feet, and gives them a very ſhewy appearance with its gloſſy hue and purple-brown color. The filaments are quite ſimple, ſtraight, rather entangled in their growth, and in length

* Synopſis of Britiſh Fuci. Intr. p. 16. & 23.

13.

I believe ſeldom exceed an inch; when young their thickneſs is regular, but with age they ſwell ſo as in ſome places to be twice as thick as in others. The diſſepiments are ſo extremely ſlender that they can only be obſerved with the higher powers of the microſcope. The joints are in length but about half equal to their thickneſs; they are nearly pellucid on each ſide towards the diſſepiments, and when the plant is old the juices collapſe into globular granules, of which three are uſually diſpoſed tranſverſely in each joint, though ſometimes a ſingle one occupies the whole. C. fuſco-purpurea approaches in many reſpects to C. Curta and Roth's C. atro purpurea, but in the latter ſpecies the juices are ſaid to collapſe into a double row of granules, and the ſize as well as the place and mode of growth are very different, and from the former it is diſtinguiſhed by the color and texture of the filaments, and ſtill more effectually by the ſhortneſs of the joints.

In drying it adheres very firmly to both glaſs and paper.

A. C. fuſco-purpurea, natural ſize.

B. filaments of ditto, magnified 1.

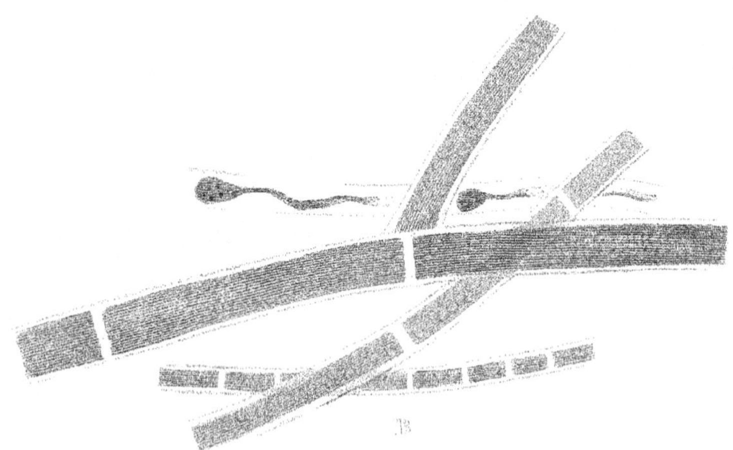


*Conferva crispata.*

# CONFERVA CRISPATA.

C. filamentis ramofis denfiffimè implicatis, crifpatis; ramis alternis remotiffimis; articulis cylindraceis longitudine diametrum multoties fuperantibus, ficcitate alternatim compreffis.

C. crifpata. Roth. Cat. Bot. I. p. 178. III. p. 275. Flora Germanica. III. pars. 1. p. 508.

In ditches and pools, about Newton Nottage, Glamorganfhire. *W. W. Young.* Alfo about London and Yarmouth.

---

I HAVE gathered C. crifpata in the neighbourhood of London and Yarmouth, and Mr. Young has brought it from the vicinity of Newton Nottage, but I do not think that it is of fuch frequent occurrence as moft of the other fpecies. It generally grows in ftagnant water, and floats in large entangled maffes on the furface. The filaments are of a dark green, wholly deftitute of glofs, and from fix or eight inches to a foot in length; they are repeatedly divided in a fomewhat dichotomous manner by alternate branches, which are always difpofed at a great diftance from each other; the joints are cylindrical, and in length many times greater than the diameter. In the older plants, the fporangium, or internal tube, which contains the granular fubftance, fuppofed by Dr. Roth to be the feeds, frequently contracts fpirally. This appearance is not however sufficiently general to authorize its introduction into the fpecific character, as Dr. Roth has done in the firft and fecond, but very properly omitted to do in the third fafciculus of his highly interefting Catalecta Botanica. When dried the joints become alternately compreffed.

The diffimilar mode of ramification, and length of the joints readily diftinguifh this fpecies from C. fracta; and from C. amphibia β, to which it bears moft refemblance, it may be at once known by its far different joints.

13.

In drying it adheres, though not firmly, to either glass or paper.

A. C. criſpata, natural ſize.

B. ditto magnified 1.

# CONFERVA FENESTRALIS.

C. filamentis repentibus minutiſſimis, tenuiſſimis, ramoſis, centrifugis: ramis plerumque divaricatis.

C. feneſtralis, Roth, Fl. Germ. III. pars I. Cat. Bot. II. p. 161. III. p. 180.

On Glaſs.

---

I OBSERVED that ſeveral of the pieces of glaſs on which I preſerve my Conſervæ, and which had lain in a damp place were covered over with a very minute mucor-like down, which on examination in the microſcope I found to accord ſo nearly with the deſcription of Roth's C. feneſtralis, that I feel no heſitation in publiſhing it as that ſpecies. The filaments are of a light grey, inclining to aſh color, and ſo minute that the glaſs on which it grows has rather the appearance of being ſoiled than covered by vegetation; it adheres to the dried conſervæ, or ſome minute ſubſtance which may moſtly be obſerved about the roots; from this as a centre numerous filaments iſſue in all directions; they are uſually from about two to four lines in length, and when they meet with any proper ſubſtance ſtrike root, and throw out other filaments in the ſame way. The branches are numerous, and generally divaricate, but the mode of ramification is very irregular, ſome of the branches being alternate, ſome oppoſite, and three or four are not unfrequently diſpoſed without interruption on the ſame ſide. Diſſepiments may be occaſionally diſtinguiſhed, dividing the filaments into joints, of which the length is generally about thrice greater than the diameter. The fructification is unknown, but may probably conſiſt in ſome granules, which are often obſervable on the branches. In drying C. feneſtralis undergoes no change. The drawing was made with the higheſt power of a compound microſcope; the extremely ſmall ſize of the filaments rendering the plant almoſt inviſible to the naked eye, and conſequently precluding the poſſibility of figuring it in its natural ſtate.

14

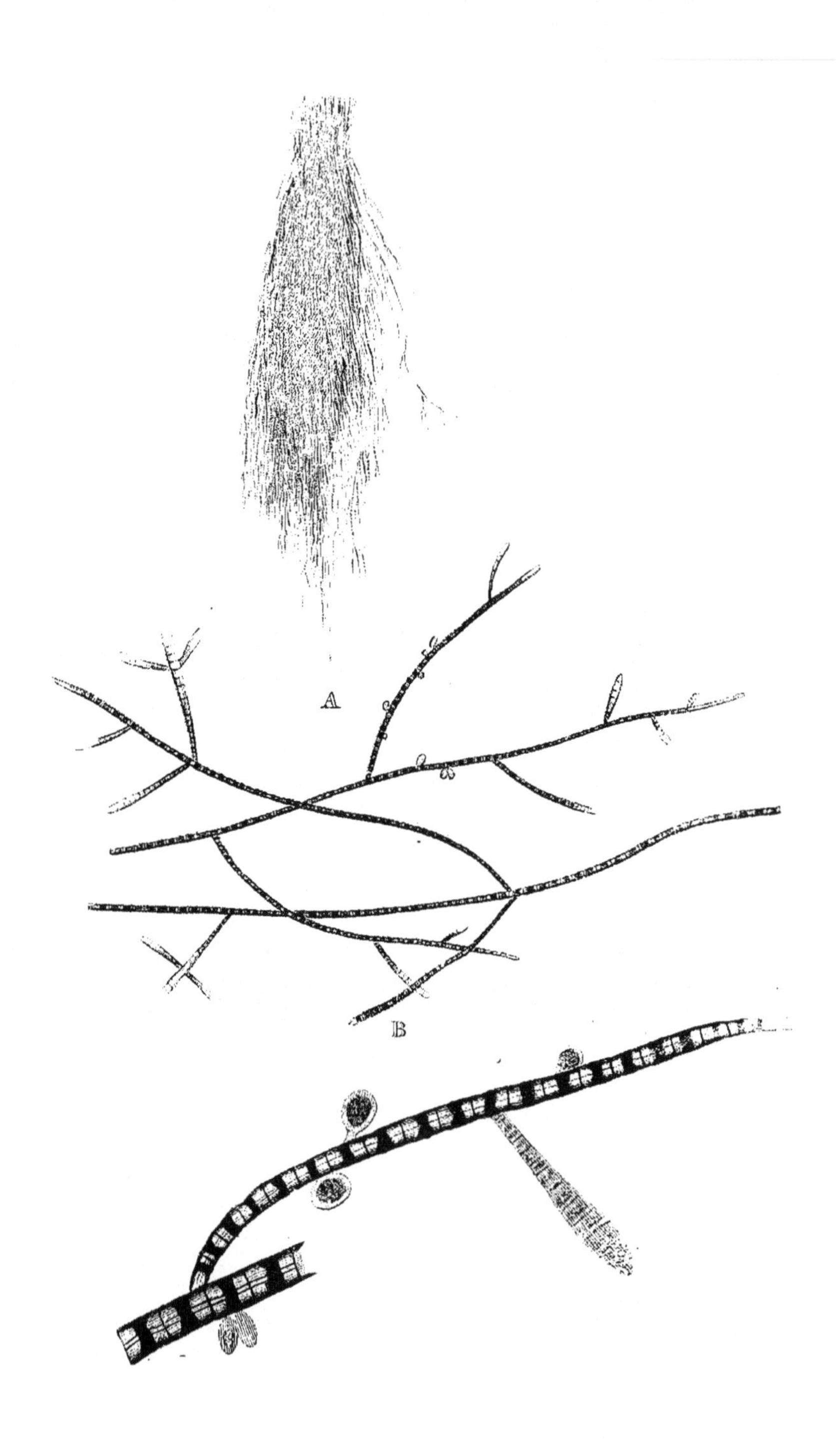


*Conferva fusca*

# CONFERVA FUSCA.

C. filamentis ramofis venofis, ramis diftantibus fub-alternis; ramulis patentibus clavatis; articulis breviufculis, medio fafciatis; capfulis fub-globofis.

C. fufca. Fl. Ang. p. 602. With. IV. p. 141.

On Rocks and Stones in the Sea. Anglefea. *Rev. Hugh Davies.* At Newton Nottage, Glamorganfhire. *W. W. Young.* At the Worms Head, and other places in the extremity of Gower.

---

I RECEIVED a fmall fpecimen of the plant here figured from my friend the Rev. Hugh Davies as the *C. fufca* of the Flora Anglica, and I conceive that this gentleman's well known accuracy, and former intimacy with Hudfon, will prove my fufficient juftification for publifhing it as fuch, more efpecially as the defcription in that Work applies better to this than to any other fpecies with which I am acquainted. I however confefs that in fo doing I feel fome hefitation arifing from the uncertainty that muft attend the elucidation of Hudfon's dark-colored marine fpecies, which has already been mentioned in the defcription of C. fucoides.

C. fufca grows in tufts from three to five inches long, and varying in color from a dull to a reddifh brown. The filaments are numerous from the fame root, and generally repeatedly branched. The branches long, remote, moft commonly alternate, and often hefet with fhort club-fhaped ramuli, which generally form a greater angle with the branches than is formed by the branches with the ftem. Mr. Young brought me a few half grown fpecimens from Newton, in which the branches were much lefs numerous than in thofe I gathered in Gower, and I believe the plant is fubject to confiderable variation in this refpect. The length of the joints but little exceeds their diameter; under the microfcope they appear of a light brown with a tranfverfe band in the middle, which nearly difappears when the juices have collapfed by drying. The capfules are globofe, rather fmall for the fize of the plant, and are fometimes raifed on fhort fruit ftalks.

In drying it does not adhere firmly to paper, and ftill lefs fo to glafs.

A. C. fufca, natural fize.
B. Ditto magnified 3.
C. Ditto ditto 1.

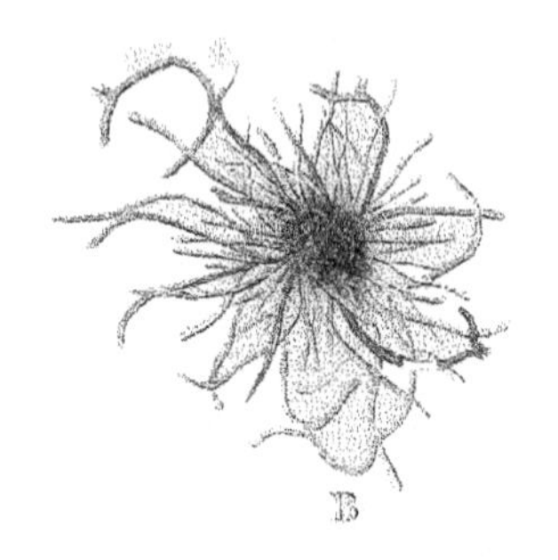

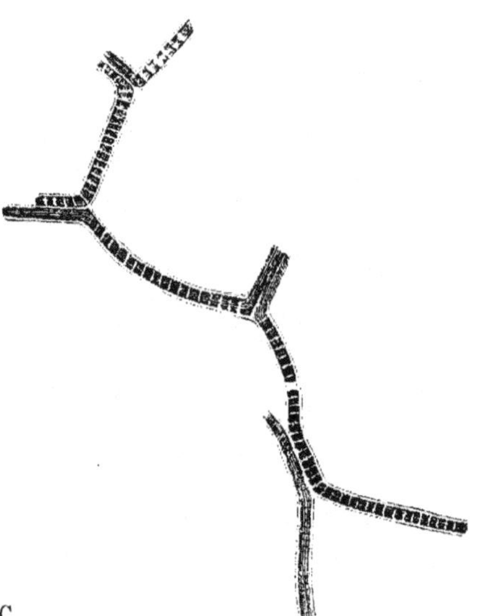

C

*Conferva mirabilis.*

Published by L.W. Dillwyn, July 1st 1808.

# CONFERVA MIRABILIS.

C. filamentis ſpurie-ramoſis, cylindricis, cœruleo-vireſcentibus; ramis e coadunitis genuflexuris filamentorum; articulis breviſſimis.

On Stones, and the Stems of Hypnum ruſcifolium in the Stream which runs through the Wood at Penllergare, near Swanſea.

---

THOSE Species to which Vaucher has given the generic name of Oſcillatoria, and which he has placed among the Tremellæ, are arranged as Confervæ by Dr. Roth, and form the diviſion 'ſporangium annulis' in the Catalecta Botanica. To this diviſion the preſent plant belongs, but it does not appear to have been heretofore deſcribed, and it differs ſo ſingularly from all its congeners as to induce me to give it the ſpecific name of *mirabilis*. I diſcovered it intermixed with C. decorticans in the above-mentioned ſtream, and alſo attached to the ſtems of Hypnum ruſcifolium, but in ſuch ſmall quantity that although I have repeatedly ſearched for it, I have not been able to obtain more than five or ſix ſpecimens.

C. mirabilis grows in ſmall thickly entangled patches, of which the diameter in the largeſt of my ſpecimens does not exceed half an inch. The color and ſize of the filaments, and the ſize and nature of the joints entirely resemble thoſe of C. diſtorta; and it is only by their different modes of growth, or with a glaſs of ſufficient power to diſcover the ſingular connection of its filaments, that it can be readily diſtinguiſhed from this ſpecies. The manner in which the filaments anaſtomoze is not ſimilar to that of jugalis, and the other ſpecies of Vaucher's genus *conjugata*, as there is no appearance of the connecting tubes, ſo ſtriking and ſingular in thoſe ſpecies. It is remarkable for having altogether the look of a branched plant, though at the ſame time it is in reality completely ſimple,

14

ſuch an appearance originating from the union of the ends of two of the filaments, each of which becomes geniculate at the beginning of the connection, and theſe ends are moſt commonly nearly of the ſame length. Other parts of the filaments are alſo frequently and ſometimes repeatedly connected with each other, in the ſame manner, and I have ſeen ſome which at firſt ſight bore a ſtriking reſemblance to a meſh of C. reticulata. I feel myſelf at a loſs even to offer a conjecture on the nature of this ſingular union of the filaments, and can only remark that they do not appear to effect any alteration in the interior of the joints, as is the caſe with C. jugalis, bipunctata and their congeners. When the juices have a little collapſed by drying, the tubular ſtructure may be readily obſerved.

C. mirabilis, in drying, adheres to both glaſs and paper.

A. C. mirabilis, natural ſize.
B. Ditto magnified 3.
C. Ditto ditto 1.

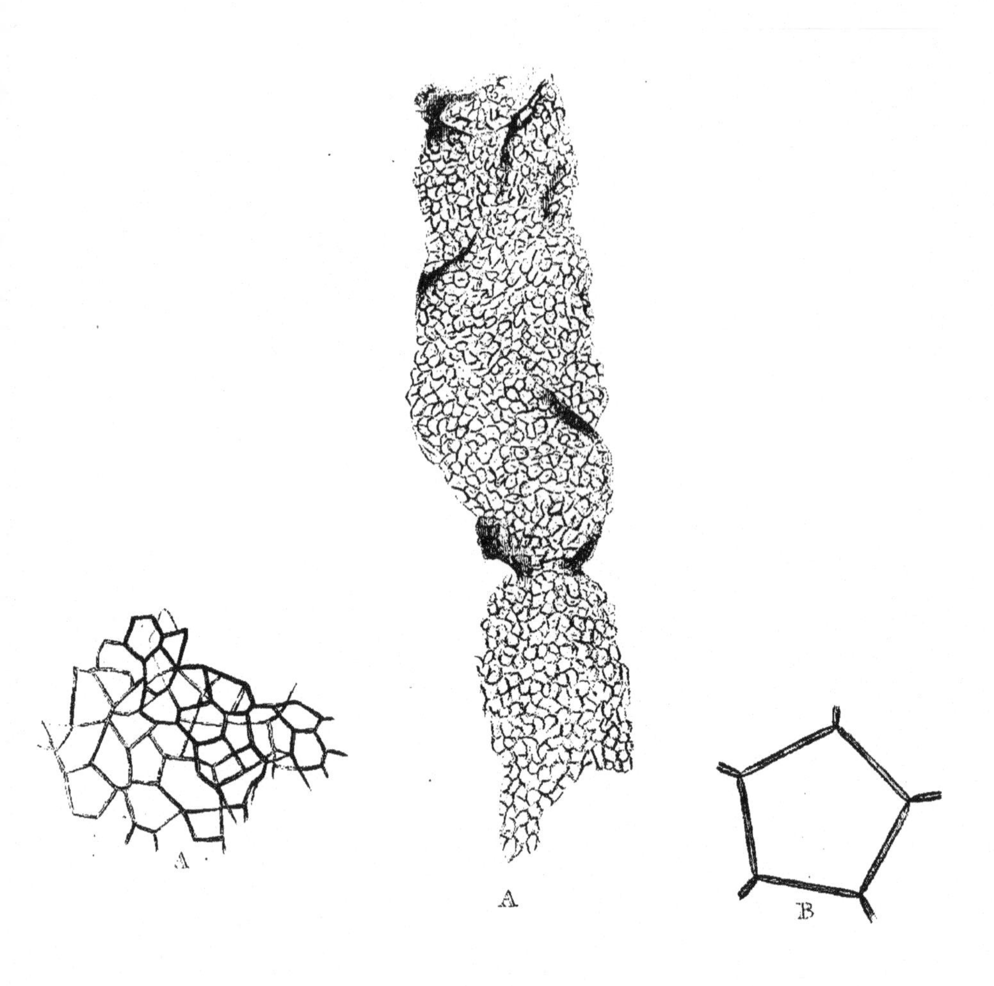


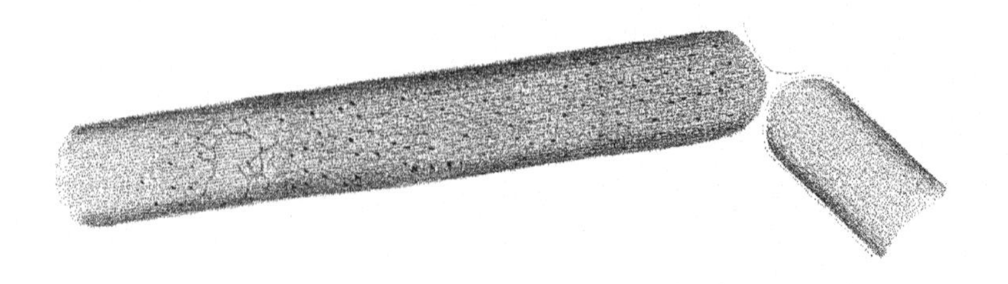

*Conferva reticulata.*

# CONFERVA RETICULATA.

C. filamentis anaſtomoſantibus, reticulatis in maculas ſub pentagonas coadunatis.

C. reticulata. Sp. Pl. p. 1635. Fl. Ang. p. 596. With. IV. p. 132. Eng. Bot. t. 1687. Ray Syn. p. 59. Dill. Muſc. p. 20. t. 4. f. 14. Hall. Hiſt. p. 2119. Pluk. Phyt. t. 24. f. 2. Morriſon Hiſt. Oxon. III. p. 644. Sec. 15. t. 4. f. 4.

Hydrodiction majus. Roth. Cat. Bot. II. p. 238.

H. tenellum. Roth. Cat. Bot. II. p. 239.

H. utriculatum. Fl. Germ. III. pars I. p. 531. Cat. Bot. III. p. 322.

H. pentagonum. Vaucher Conferves d'eau douce. p. 88. pl. 9.

In Ditches and Ponds, about Hounſlow, *Hudſon*. In the Cam and Pool of the Botanic Garden at Cambridge. *Relhan*. Heigham, near Norwich, *Mr. Pitchford*. In the Stream near Low Hall, Walthamſtow. *E. Forſter, jun.* Ditches at Woburn and Apſley. *Dr. Abbot.* Ditches at New Hall, near Kenfield, Suſſex. *Mr. Borrer.* Thorpe, near Norwich. *Mr. S. Wilkin.*

---

ON the ſame morning I received freſh ſpecimens of the preſent ſingular ſpecies from my friends the Rev. J. Davies and W. J. Hooker. The former gathered it in the Pool of the Botanic Garden at Cambridge, and the latter ſent it me from the neighbourhood of Norwich, where it was gathered by Mr. S. Wilkin. It floats in irregular maſſes on the ſurface of Ponds and Ditches, and though it has been diſcovered here and there in different parts of this kingdom and of the Continent, yet its known *loci natales* are comparatively ſo few that it muſt be reckoned among the rareſt of the freſh water Confervæ, as the ſpecies has been long well known to botaniſts, and the ſingularity of its ſtructure precludes the poſſibility of its ever having been confounded with any other. The whole plant forms an oblong net-like tube, varying from a few inches to a foot in length, and from half an inch to two inches in diameter; being all formed of meſhes which are moſt uſually pentagonal, but ſome are compoſed of four and ſome of fix ſides. Each ſide is formed by a ſingle joint which branches in a dichotomous and almoſt divaricated manner at each end, ſo that theſe branches aſſiſt in forming other meſhes. The joints are cylindrical, and vary greatly in the ſame plant, ſome being twice as large as others, and the breadth varies proportionably from the ſize of human to that of the thickeſt horſe hair. Reſpecting the propagation of this ſpecies I cannot do better than copy the result of M. Vaucher's observations; they are ſo extremely curious and intereſting that I earneſtly recommend them to the notice of ſuch botaniſts as reſide in the neighourhood in which the plant grows, and ſhall only add that I long heſitated to give them credit, but confeſs that the few obſervations which my opportunities have allowed me to make tend ſtrongly to confirm them.

" Enfin, le 24 Germinal j'arrivai à ce but tant déſiré, et je vis d'un ſeul coup d'œil, toute la reproduction de l'hydrodictye. Chacun des cinq filets qui forment le pentagone commença à se renfler légérement, ſur tout à ſes extrémités. Enſuite il s'en ſépara, non pas par une rupture proprement dite, mais en ſortant de l'intérieur de la membrane dans laquelle il etait contenu, et qui ſans doute s'etait ouverte; et après cette ſéparation, il flotta dans l'eau ſous la forme d'un bâton cylindrique. Bientôt il s'aplatit, et éprouva un altération que je comparerai à celle qu'nn commencement de fuſion produit ſur les métaux; ensuite il s'agrandit inſenſiblement dans tous les ſens, et les mailles dont la réunion le conſtituait s'etant écartées les unes des autres, il devint lui même un nouveau réſeau que l'on diſtinguait au microſcope. Bientôt ces mailles purent être obſervées à la vue ſimple, et enſin chaque bâton fut totalement changé en un réſeau entièrement ſemblable à celui dont il faiſait partie. Toutes ces tranſformations s'opérerent dans l'eſpace de quelques jours, et au bout de deux ou trois mois les jeunes réſeaux avaient acquis toutes les dimenſions dont ils étaient fufceptibles. Quoique je n'éuffe aucune doute fur ce mode de reproduction, je n'ai pas laiffé de le fuivre pendant les deux années qui fe font ecoulées depuis ma première obfervation. J'ai donc vu ces réfeaux qui étaient n'és dans l'an VIII. fe conferver pendant tout l'été fans reproductions nouvelles et enfuite de développer au printemps de l'an IX, comme les autres s'étaient développés l'annéc précédente, et au moment où j'écris (1re Floréal, an X,) quoique le printemps ait été extraordinairement fec, et que le foffé où vit l'hydrodictye foit entirement privé d'eau les filets que j'y ai recueilles, et que j'ai rapportés chez moi ne font pas moins développés comme les autres années. Voilà donc une exemple d'emboîtement peut-être plus remarquable que tous ceux qui, jufqu'à préfent ont été obfervés. En effet il n'est guères permis de mettre en doute que fi les côtés des mailles du réfeau de l'année précédente, étaient les réfeaux de cette année, les côtés des mailles des réfeaux actuels font auffi les réfeaux de l'annéc prochaine, que chaque fibre de ces mailles est ellemême le réfeau qui fe developpera dans deux ans, et que chaque fibrille de la fibre principale fera le réfeau qui fe développera dans trois ans, et ainfi de fuite, jufqu'à ce qu'il plaife à l'auteur de la nature de mettre fin à ce développement en détruifant l'efpèce qui le préfente."

In drying, C. reticulata adheres, though not very firmly, to either glafs or paper.

A. A. C. reticulata, natural fize.
B. A mefh of ditto magnified 5.
C. A joint of ditto ditto 1.

# CONFERVA CORALLINA.

C. filamentis, ramofis, dichotomis, lubricis; diffeminentis contractis, articulis furfum incraffatis, longis, fub-pyriformis; fructu involucro polyphyllo fubtenfo.

C. corallina, Fl. Scot. p. 988. With. IV. p. 136. Roth. Cat. Bot. III. p. 225. Eng. Bot. t. 1815.

C. corallinoides, Sp. Pl. p. 1636. Fl. Ang. p. 598.

C. geniculata. Ellis in Phil. Trauf. LVII. p. 425 t. 18. *f.* f. F.

C. marina gelatinofa, corallinæ inftar geniculata craffior. Dill. Hift. Mufc. p. 33. t. 6. f. 36.

Corallina confervoides gelatinofa alba, geniculis craffiufculis pellucidis. Dill. in Ray. Syn. p. 34.

On rocks and ftones in the fea. At Cockbufh, Suffex, and Ynys y Moch, near Bangor, *Dillenius.* Brighton, *Ellis.* Anglefea, *Rev. H. Davis.* Weymouth, *Mr. Stackhoufe.* Hartley, Northumberland, *Mr. Winch.* Cowes in the Ifle of Wight.

---

THIS beautiful fpecies may be found on feveral parts of our coafts during the fummer months, but is by no means of general growth. When not expanded in water it forms a gelatinous flippery mafs. The color when the plant is in perfection is a bright tranfparent pink, not unmixed with fcarlet, but with age or expofure to the fun becomes lighter, and often tinged with green. The root is fibrous, and throws out many filaments, which are repeatedly divided with regular dichotomies, and vary in length from three to fix inches. In proportion to the length the filaments are thicker than in any of its congeners with which I am acquainted, being about half a line in diameter. The joints, as Dr. Smith obferves, are nearly pear-fhaped, being much fwollen towards the apex, and about thrice as long as broad. Mr. Borrer, whofe opinion in all matters relating to cryptogamous plants is entitled to great deference, informs me that the fructification " confifts of a mafs of feeds not enclofed in any membranous capfule whatever, but immerfed in a jelly, fometimes forming a whirl in the contractions of the filament, and fometimes a lateral knob in the fame fituation."* The fpecimens which I gathered at Cowes produced an abundance of both whirled and lateral fruit, but the refult of a long inveftigation which I gave them, differed widely from the foregoing. The lateral knobs appeared to be perfect capfules, round all of which a tranfparent limbus was readily obfervable, and I faw feeds efcape from the apex of one precifely as in C. rubra, and the generality of the marine fpecies. In fome plants the capfules feemed to be compofed of three or more cells, refembling thofe of Euphorbia, and I thought I obferved fome of the whirls to be formed by a number of fimilar cells difpofed round the diffepiments, and thus forming a kind of polylocular capfule. Though the fhape and appearance of thefe whirls differed materially from each other, I never doubted that they were true capfules till I received Mr. Borrer's letter; and I am certain that a well defined pellucid limbus furrounded all that I examined, though frequently the feeds and mucus which had efcaped fo adhered to the outfide of the capfule as almoft to cover it, and had I not been well acquainted with that gentleman's accuracy, I fhould have imagined that this circumftance had deceived and induced him to believe that no feed veffel exifted. The whirled and lateral fructification are fometimes, though not generally found on the fame plant, and both are always fubtended by an involucrum of feveral obtufe, jointlefs, incurved leaves. This production of different kinds of fruit is far from being confined to C. corallina. I am informed that three have been difcovered on C. fetucea, which is nearly allied to the prefent fpecies, and Mr. Borrer remarks, " I fhould not be furprized by any variety of fructification in the marine algæ, having myfelf found on Fucus pinaftroides no lefs than four kinds." Refpecting the red granules

* See Eng. Bot. t. 1815.

which I have above called feeds, he adds, " In fome fpecimens they were oval, and in appearance folid; in others globofe, and feemingly divided into three, and unlefs I am very much miftaken, each feed in either cafe had a pellucid limbus." Mr. Hooker, who examined fome fpecimens which were nearly frefh, could not difcover any limbus, and he is of opinion that the feeds (commonly fo called) of both fuci and confervæ, which have a limbus, are in fact capfules. I confefs that the microfcope I ufed in the Ifle of Wight, was neither fo good or convenient as that which I commonly ufe, and therefore I much more doubt my own correctnefs than Mr. Borrer's, more especially as he has had frequent and much better opportunities of ftudying this fpecies than myfelf. My obfervations as before related afforded me no room to doubt, that the nature of both the whirled and lateral fructification is fimilar to that of Roth's Ceramia; but if Mr. Borrer's obfervations and Mr. Hooker's ideas are correct, the fructification confifts of minute capfules immerfed in a loofe tranfparent jelly, without any cafe or covering.

In drying C. corallina lofes much of its color, and adheres firmly to either glafs and paper.

A. C. corallina, natural fize.
B. D° with lateral fruit magnified 4.
C. D. D° d° 2.
E. whirled fruit d° 2.

N. B. C. D. and E. were completed from frefh fpecimens in the Ifle of Wight.

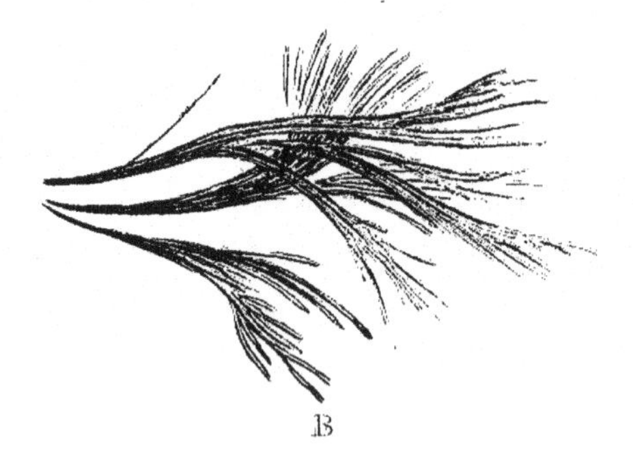


D

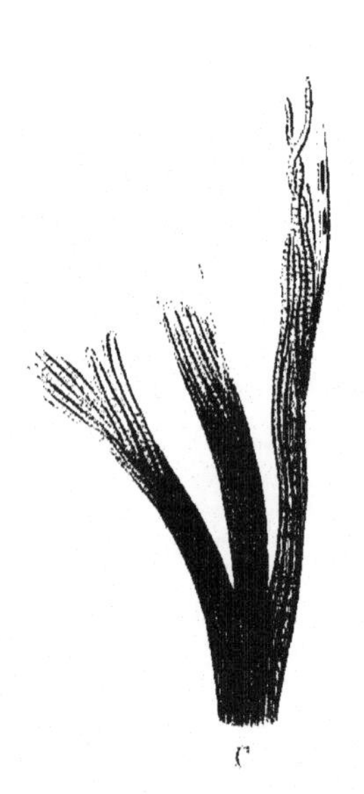


*Conferva vaginata.*

# CONFERVA VAGINATA.

C. filamentis ramofis cylindricis geniculatis cœruleo viridefcentibus, ramis vaginato-fafciculatis, articulis breviffimis

Ofcillatoria vaginata. Vaucher Hiftoire des Conferves d'eau douce. P. 200. tab. 15. f. 13.

C. velutina β. Roth. Cat. Bot. III. p. 200?

Frequent during the Winter months on damp foil, by the fides of paths, &c. about Weft Town, Suffex. *Mr. Borrer.* On Earth in the Flower-pots in a Green-houfe at Norwich. *Mr. Hooker.* On Rocks and Stones in the Stream which runs through the Wood at Penllergare, near Swanfea.

---

I DISCOVERED this fingular fpecies, growing mixed with C. decorticans, on ftones which are occafionally overflowed by the ftream, and alfo entangled among the filaments of C. fluviatilis, and the leaves of fontinalis antipyratica, in the neighbourhood of Penllergare. My friends Mr. Hooker and Mr. Borrer inform me, they alfo have found it in their refpective neighbourhoods, but in fituations fo diffimilar, that the plant feems to poffefs a perfect indifference with refpect to the foil or place in which it grows.

Though Vaucher in the drawing of this plant has not availed himfelf of the higher powers of the microfcope, his figure and defcription are too clear to admit any doubt of its being his Ofcillatoria vaginata. It may be well to remark that this author has formed limofa, fontinalis, and their congeners with fhort annular joints into a feparate genus, which he has placed among the Tremellæ, and given it the name of Ofcillatoria, from a fpontaneous motion that he fuppofes them to poffefs. Of the nature of this motion I have already hazarded an opinion in the defcription of C. limofa, and these plants, both in ftructure and

14

appearance, ſo entirely accord with the Conſervæ, that I confeſs myſelf ſur-
priſed at their having been removed by that able author to the Tremellæ, to
which they ſeem to bear a far leſs affinity. As Dr. Roth ſays he could not
diſcover any ſheath in his variety of C. velutina, to which he refers Vaucher's
C. vaginata, I have thought it right to quote it as a ſynonym with a mark of
doubt.

C. vaginata grows in ſmall tufts, of which the diameter of the largeſt that I
have gathered does not much exceed a quarter of an inch, and the greater part of
them are in fragments of a ſtill ſmaller ſize. The filaments are cylindrical, and
reſemble thoſe of C. limoſa, except that they are branched, and that they are
encloſed in bundles within a membranous ſheath, which is ſo peculiar to this
ſpecies that it is alone ſufficient to diſtinguiſh it from every other I am acquainted with: theſe ſheaths are themſelves branched or divided repeatedly into
ſmaller ones, at irregular diſtances of various ſizes; they are narroweſt at their
origin, and become ſwollen upwards, as the filaments increaſe by branching, ſo
as ſometimes to reſemble a ſeries of Cornucopiæ. The ends of the filaments
which are of various lengths project beyond the ultimate diviſion of the ſheath,
and they are ſometimes curiouſly coiled round each other. It appears probable
that this ſpecies is propagated by the ſeparation of the different diviſions of the
ſheath, each of which may thus form a diſtinct and perfect plant, and Vaucher
goes ſo far as to ſuppoſe that every individual filament at length becomes an
envelope for other filaments which are generated within them.

In drying, C. vaginata adheres, though not firmly, to either glaſs or paper, and
when dried, may be revived by immerſing it in water.

A. C. vaginata, natural ſize.

B. Ditto magnified 3.

C. Piece of ditto, magnified 1.

D. Ditto, larger than it appeared in the microſcope.

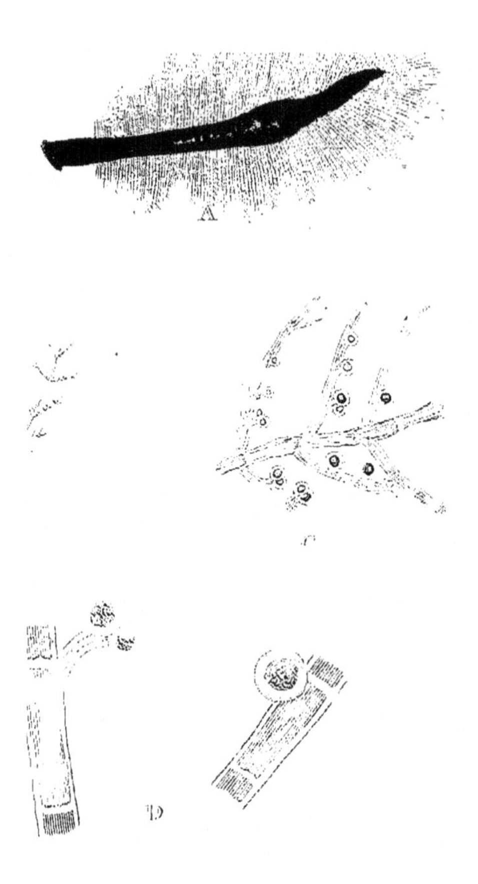


*Conferva Turneri.*

# CONFERVA TURNERI.

C. filamentis pinnatis fafciculatis; primis oppofitis fub-fimplicibus; articulis longis diffepimentis pellucidis; capfulis in pinnis infra medium fecundis, pedunculatis, globofis.

Ceramium Turneri. Roth. Cat. Bot. III. p. 128. Tab. 5.

On Fuci and Corallines in the Sea at Cromer. *D. Turner, Efq.*

---

THE prefent fpecies was firft difcovered fome years ago by my friend Mr. Turner, and was communicated by him to Dr. Roth, who named it in honor of its difcoverer, and publifhed a defcription in the third Fafciculus of his Catalecta Botanica, with a good drawing from the accurate pencil of Profeffor Mertens. The fpecies which has fubfequently been figured in Englifh Botany under the fame name, is the C. plumula of Ellis and of this work.

C. turneri is found in great abundance on fuci and corallines in the fea at Cromer, during the fummer months, and from its elegant growth and delicate rofe color, may be confidered one of the moft beautiful of the Confervæ. Its habit is bufhy, forming thick tufts. The filaments rarely exceed an inch in length, and are undivided, but hefet with oppofite and moftly fimple pinnæ, from four to fix lines long, between patent and horizontal, which are fufficient readily to diftinguifh this fpecies from C. rofea, to which in appearance it is moft allied. The length of the joints is about thrice greater than their diameter, and they are perfectly colorlefs at their diffepiments. The capfules are numerous, globofe, moftly raifed on fhort footftalks, and arranged together on the upper fide of the lower pinnæ: though in general folitary, it occafionally happens that two are fupported on the fame peduncle.

15

For the drawing I am obliged to my ſriend William Jackſon Hooker, Eſq. to whom for many other valuable communications this work is alſo greatly indebted.

In drying it adheres to both Glaſs and Paper.

| | | | |
|---|---|---|---|
| A. | C. turneri, | natural ſize. | |
| B. | Ditto, | magnified | 4. |
| C. | Ditto, | ditto | 3. |
| D. | Ditto, | ditto | 1. |

A

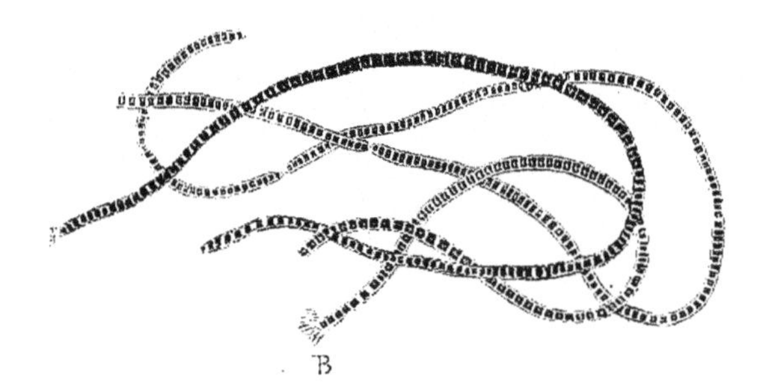

B

*Conferva atro-purpurea!*

Published by L. W. Dillwyn, Feb.y 2nd 1809.

# CONFERVA ATRO-PURPUREA.

C. filamentis ſimplicibus, ætate hic illic inæqualiter toroſis, atro-purpureis; articulis diametro dimidio brevioribus, ſingulis ſeriem dupliciem globulorum includentibus.

C. atro-purpurea. Roth, Cat. Bot. III. p. 208. t. 6.

Bantry Bay, Ireland. *Miſs Hutchins.*

---

C. atro-purpurea was firſt diſcovered growing upon mill-wheels, in the vicinity of Bremen, and communicated to Dr. Roth by Profeſſor Mertens. Miſs Hutchins has lately gathered it in Bantry Bay, and from her, through the medium of our mutual friend, Mr. Turner, I have received ſpecimens of this, as well as of ſeveral other ſpecies at preſent undeſcribed; an account of which I ſhould have been happy to publiſh, had they not ſuffered too much change in drying. The preſent is one of the few Confervæ that may be reſtored by immerſion in water, and I have therefore ventured to make the annexed drawing from a dried ſpecimen.

The root is fibrous; the filaments grow in ſmall tufts, they are about two or three inches in length, thinner than human hair, nearly ſtraight, of a gloſſy hue, and dark purple color. As in C. fuſco purpurea, when the plant is young the filaments are moſt probably of an uniform thickneſs, and they are ſo deſcribed by Dr. Roth, but thoſe now before me are in ſome parts ſwelled, and much thicker than in others; the diſſepiments are narrow and pellucid; the joints are in length but about half equal to their diameter, and each contains two rows of granules diſpoſed tranſverſely, which, like thoſe of C. bipunctata, occaſionally take a ſtellated appearance. A longitudinal pellucid line is obſervable

15

running through the middle of ſome filaments, and in others the bands of granules are divided in like manner into three or four ſeparate compartments.

C. atro-purpurea is very cloſely allied with C. fuſco-purpurea, but in that ſpecies there is only a ſingle band of granules in each joint.

In drying it adheres to both Glaſs and Paper.

A. C. atro-purpurea, natural ſize.
B. Ditto, magnified 2.
C. Ditto, ditto 1.

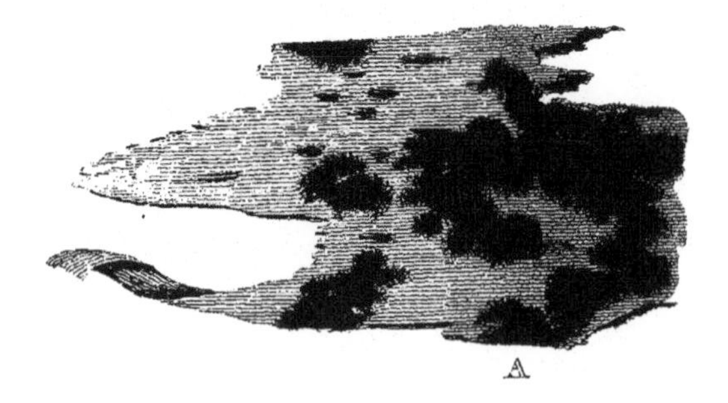

A

B

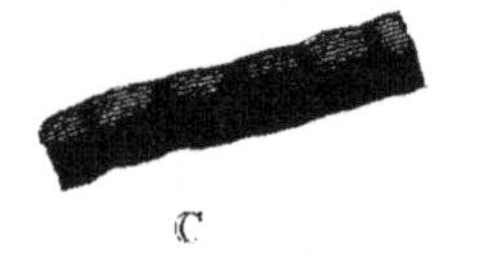

C

*Conferva ebenea.*

ker Esqr. delt.

Published by L.W. Dillw -

# CONFERVA EBENEA.

C. filamentis ramosis, erectis, cæspitosis, brevibus, rigidis, sub-cartilageneis; ramis ramulisque obtusis; articulis diametrum longitudine æquantibus; dissepimentis contractis.

Conferva nigra. Roth, Cat. Bot. III. p. 299.

Byssus nigra. Fl. Ang. p. 606. Fl. Scot. p. 1003. With. IV. p. 144. Fl. Germ. III. pars 1. p. 567.

Byssus petrea nigerrima fibrosa. Dill. Hist. Musc. p. 9. t. 1. f. 18.

Byssus minima saxatilis nigra ramosissima, &c. Micheli. Gen. Plant. p. 212. Tab. 90. f. 5. & Byssus cæspitosa nigra, &c. p. 211. Tab. 90. f. 7. ejusd. lib.

Byssus nigra velutina. Hall. Hist. p. 2104.

On Rocks and Trees. *Dillenius.*—On Rocks in the Highlands. *James Brodie, Esq.*—On the Stump of a Tree in Mackbeth's Wood, at Brodie, near Forres, N. B. *W. J. Hooker, Esq.*—On Birch Trees, at Costesy near Norwich. *Mr S. Wilkins.*

---

C. EBENEA, accompanied by the accurate drawing, which is represented in the annexed plate, was obligingly communicated to me by my friend Mr. Hooker, who, in company with Mr. Turner, gathered it near Forres, in Scotland. Authentic specimens with which I have been favored by Sir Thomas Frankland and the Rev. Hugh Davies, prove that Hudson's *Conferva nigra,* respecting which I had previously been accustomed to yield to the generally received opinion of its being the same as *Fucus fruticulosus,* is in reality the C. *atro rubescens* of this work. In consequence of this it became necessary to change the name given to the species here figured by Dr. Roth, who in the third Fasciculus of his Catalecta Botanica has, with great propriety, removed the plant from the Byssi to the Conferva, but has retained the specific name of

15

the Flora Anglica, in the place of which I have adopted a nearly ſimilar appellation, propoſed by my friend Sir Thomas Frankland.

C. ebenea grows on rocks and trees in thick black tufts, together forming patches of various ſizes, but it is not by any means a common ſpecies. Mr. Turner tells me that at a little diſtance the patches look like ſmall ſpots of ſoot. The filaments I believe never exceed three or four lines, and are moſt frequently conſiderably leſs than a line in length; their ſubſtance is ſtiff, ſomewhat horny, and their growth erect: they are about twice branched in a ſub-dichotomous manner, and the branches are irregularly beſet with ſimple patent ramuli with obtuſe apices. The diſſepiments are opake, more or leſs contracted, and divide the filaments into joints, of which the length about equals their thickneſs. No fructification has been diſcovered.

In drying it adheres but very ſlightly to either Glaſs or Paper.

A. C. ebenea, natural ſize.
B. Ditto, magnified 3.
C. Ditto, ditto 1.

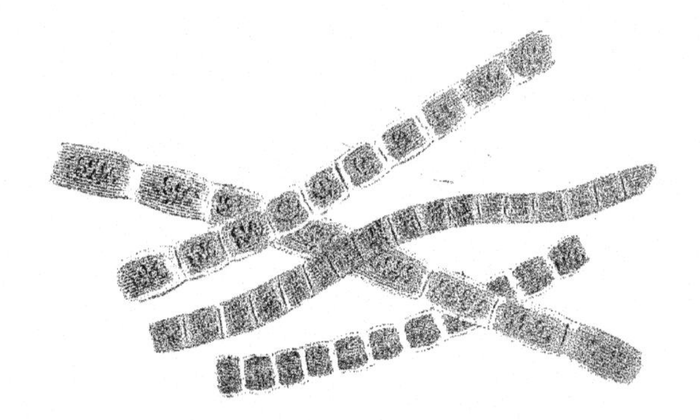

*Conferva Youngana*

J. Simpkins sculp.

Published by L.W. Dillwyn

# CONFERVA YOUNGANA.

C. filamentis cæſpitoſis, ſimplicibus, rigidiuſculis, apicibus obtúſis; diſſepimentis contractis; articulis breviuſculis, adultioribus ſubnodoſis; ſuccus in globulos ſolitarios demum congeſtus.

On the Lime-Stone Rocks near Dunraven Caſtle, Glamorganſhire. *W. W. Young.*—On the Piles of the Jetty at Great Yarmouth, and Cromer, Norfolk. *W. J. Hooker*, Eſq.

---

THE preſent ſpecies was firſt diſcovered by Mr. William Weſton Young, A.L.S. in honor of whom I have named it, as a token of my private friendſhip, and as a public acknowledgment of the aſſiſtance which this work has received from his accurate pencil.

C. youngana grows very plentifully on the limeſtone rocks about Dunraven, frequently in ſuch places as are never covered by the ſea, and only waſhed by the ſpray at high water; and Mr. Young tells me that it never grows much lower than high water mark, or where it is not left expoſed to the air during the greater part of the day. The ſituations in which Meſſrs. Turner and Hooker have found it at Yarmouth and Cromer are in this reſpect ſimilar. It forms elegant little tufts, uſually about a quarter of an inch in length, and of a dark green color. The filaments are ſimple, ſomewhat rigid, obtuſe at the apices, and when the plant is at maturity they become contracted at each diſſepiment. The length of the joint varies conſiderably in different filaments, being ſometimes only equal to and at others double the diameter. In the young plants theſe joints are nearly of the ſame color throughout, but with age they become more pellucid towards the diſſepiments, and at length the green matter collapſes into a globule which ſometimes diſappears, and leaves the filaments perfectly colorless.

15

In drying C, youngana adheres to Paper, but not at all firmly to Glafs.

A. C. youngana, natural fize.

B. Ditto, magnified 1.

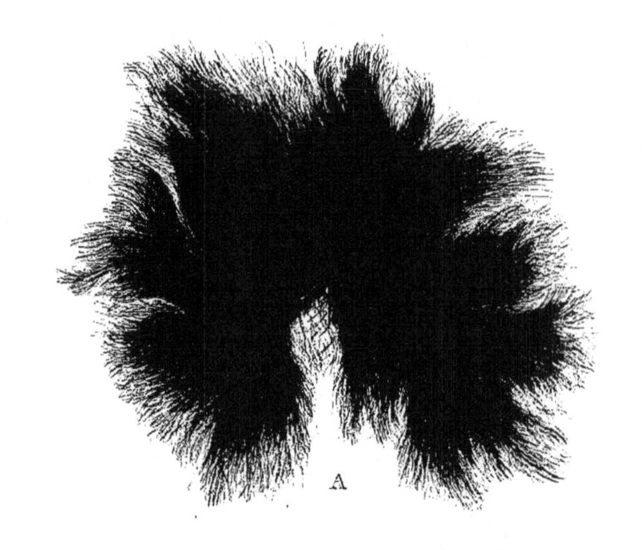


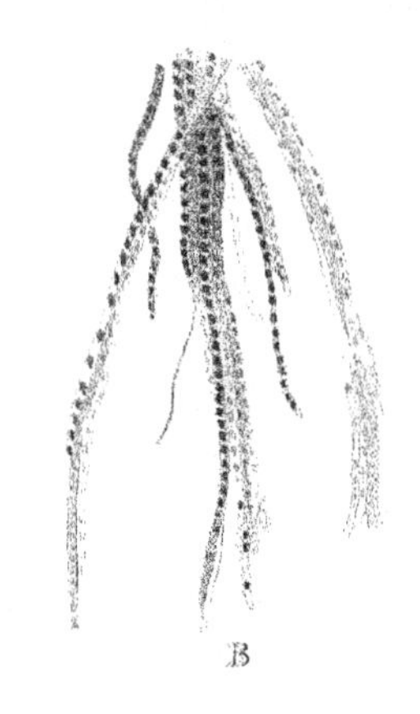


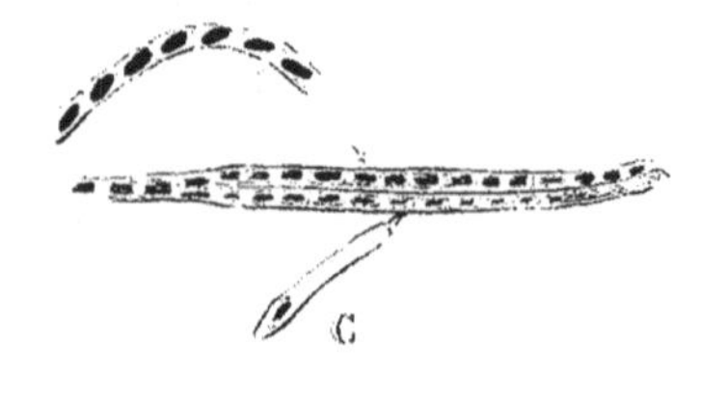


*Conferva fœtida.*

# CONFERVA FŒTIDA.

C. filamentis ramofis, flaccidis, virgatis, coadunatis, apicibus liberis; ramis confertis fub-dichotomis; diffepimentis obfoletis, articulis longiufculis granula elliptica folitaria includentibus.

Ulva fœtida, Vaucher. Hiftoire des Conferves d'eau douce. p. 244. t. 17. f. 3.

Stagnant Pools in the Salt Marfhes at Cley, Norfolk; *Mr. Hooker.* Bantry Bay; *Mifs Hutchins.* Among the Rocks near low water mark, under the Mumbles Light-Houfe, Glamorganfhire.

---

IN the early part of laft June I difcovered this curious production of nature, growing under the Mumbles Light-Houfe, in a pool left by the tide, near low water mark, where, had not the tide receded unufually low, it would not have been expofed to view. This I at firft fuppofed to be its natural fituation, and the caufe of its not having been previoufly difcovered, but I have fince learnt that Mr. Hooker had gathered it two months before, in the falt marfhes above mentioned, and had afcertained it to be the plant defcribed and figured by Vaucher. C. fœtida, therefore, feems to poffefs an unufual indifference with refpect to its place of growth, for, he fays, "Elle fe rencontre dans touts les eaux fraiches et courantes des petits ruiffeaux." I have not ventured on introducing it as a vegetable without confiderable hefitation, on account of its ftrong peculiar oily * fmell, refembling that of fome of the zoophites, but the eye, even when affifted with the higheft powers of a microfcope, cannot difcover any

* The remark made by Vaucher upon the fmell of this plant, agrees almoft exactly with what I had obferved before I had any idea of my plant being the fame as his. He fays, "L'odeur qu'elle repand est très forte, et reffemble aux odeurs animales et furtout à celle des corps qui commencent à entrer en putréfaction."

appearance at all ſufficient to diſtinguiſh it from the tribe with which it is now arranged.

C. fœtida grows in thick buſhy tufts, near two inches in length and of a dull olive color. At firſt ſight it very much reſembles C. littoralis, but when examined under a glaſs it differs entirely from this and every other ſpecies with which I am acquainted. The root appears to be a very minute callus, from which numerous ſhort creepers are thrown out, but it is ſo ſmall as to be hardly obſervable. The filaments are very flaccid, and peculiarly ſlender in proportion to their length; they are twice or thrice branched in an irregularly dichotomous manner, and in their adheſion to each other reſemble thoſe of C. vaginata, but there is not any appearance of a ſheath. The branches at their baſe, and frequently through nearly their whole length are cloſely united to the ſtem, in the ſame manner as are the main filaments to each other, being ſeparated only at the extremities, which gave cauſe to Vaucher's making it a part of the ſpecific character, "extremitatibus multoties diviſis". The length of the joints is nearly double the diameter, each joint contains an egg-ſhaped maſs, reſembling thoſe of C. jugalis, which, from analogy, I ſuppoſe are formed by a collapſion of their juices, or internal granules, and are ſomehow connected with the fructification, as ſuppoſed by Vaucher, but like him I have had no opportunity of inveſtigating the matter.

Villars's C. fœtida may poſſibly be the ſame plant as is here figured, but neither from his deſcription nor his figure is it poſſible to decide upon the ſubject, and I have therefore not quoted him.

This ſpecies adheres to both Glaſs and Paper.

A. C. fœtida, natural ſize.
B. Ditto, magnified 2.
C. Ditto, ditto 1.

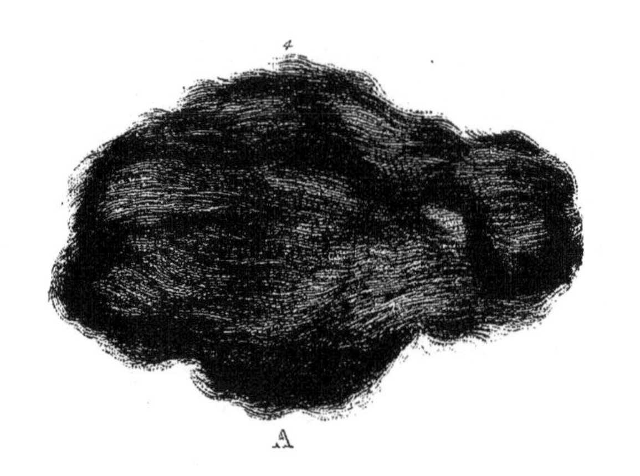
4
A

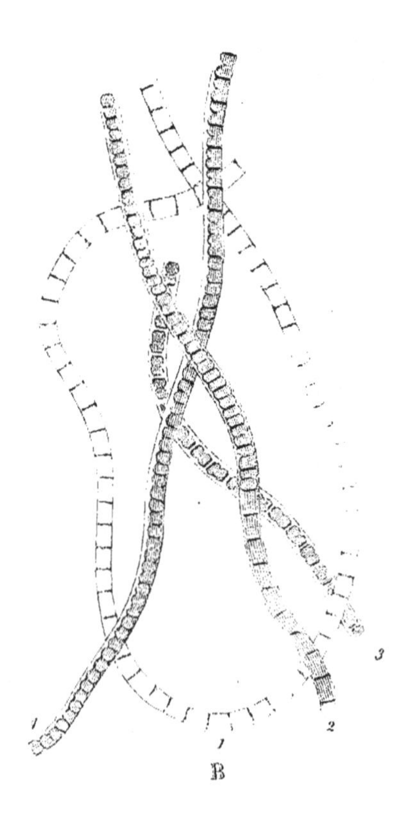
1
1
2
3
B

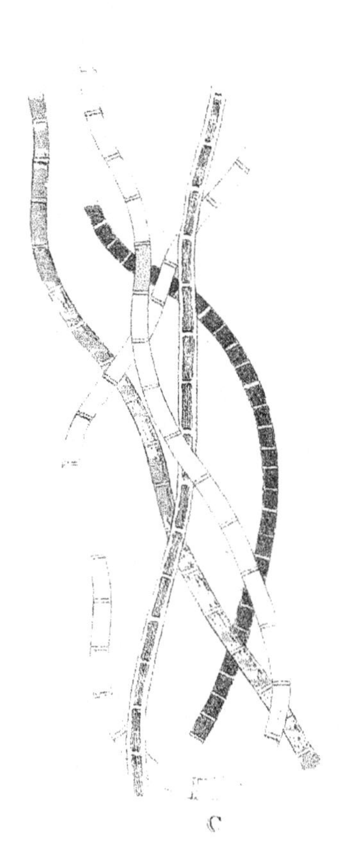
C

# CONFERVA BIPARTITA.

C. filamentis ſimplicibus, tenuibus, longiſſimis, denſiſſimè compactis, flavo virentibus; articulis diametro ſub triplo longioribus denium bipartitis.

In ſmall Pools on the Bogs on Town Hill Common, near Southampton. *Mr Woods.*

In the Ditches between Pontardylais and the Sea, Carmarthenſhire.

---

THE preſent ſpecies almoſt fills the ditches in the marſhes between Pontardylais and the ſea, and I cannot find that it has been heretofore deſcribed. It floats on the ſurface of the water in large denſely matted maſſes, of a yellowiſh green color, and retains air bubbles in the ſame manner as thoſe ſpecies which were formerly confounded together under the name of C. bulloſa. The filaments are very long, unbranched, and in thickneſs rather exceed thoſe of C. rivularis. The length of the joints is uſually from three to four times their diameter. At a certain age the interior of each joint ſeparates by a tranſverſe diviſion in the middle, into two veſicles, which at length contract and become rounded at the corners. In figure B, the filament marked No. 1 is in its youngeſt ſtate: in the lower part of No. 2, the tranſverſe ſeparation of the joints has juſt commenced, and it is ſeen in the different ſtages of advancement towards the upper end of the ſame filament and in No. 3. Theſe internal veſicles when thus contracted are ſometimes diſpoſed, as is repreſented at No. 4, and it frequently happens that the diviſion has commenced at one ſide of the filament and not on the other. The plant figured at C grew in the ſame place and manner with the foregoing, and could only be diſtinguiſhed by the microſcope. I found the filaments of both mixed with each other, and the joints of many were ſo intermediate as to prove that both belong to the ſame ſpecies.

15

C. bipartita may be diſtinguiſhed from C. ſordida by the ſmaller and remarkably pellucid filaments of the latter, as well as by their more ſimple internal ſtructure. From C. rivularis it may be known by its different color and mode of growth, and by its longer joints with two veſicles in each. Both theſe ſpecies however vary, and occaſionally approach each other in a ſurpriſing manner, and no other Confervæ have ever puzzled me ſo much. I have gathered C. rivularis, in ſome of the filaments of which there has been a pellucid line running longitudinally through them, as if they were about to ſeparate in that direction. It alſo frequently happens that the coloring matter in the joints of that ſpecies is collapſed alternately on both ſides of the filament, ſo as to preſent a curious zic-zac appearance, and I once ſaw the internal veſicles of C. bipartita arranged in the ſame manner.

In drying it adheres, though not ſo firmly as C. rivularis, to either Glaſs or Paper.

A. C. bipartita, natural ſize.

B. Ditto, magnified 1.

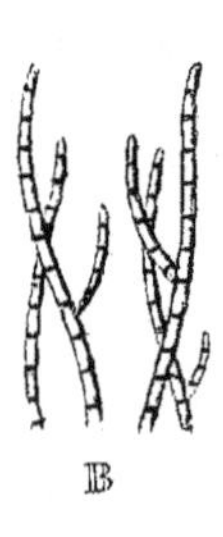

*Conferva Acharii.*

J. Simmons

Published by L. W. Dillwyn

# CONFERVA ACHARII.

C. filamentis ramoſis, cæſpitoſis, rigidiuſculis, ſub-erectis, fragilibus, fuſco-olivaceis; ramis brevibus, patentibus, apicibus obtuſis; articulis longiuſculis.

C. Acharii. Weber & Mohr, reiſe durch Schweden, p. 104. t. 1. f. 6. Roth. Cat. Bot. III. p. 298.

Parmelia velutina. Acharius Methodus Lichenum. II. p. 245.

On ſhady Banks in the neighbourhood of Norwich, not uncommon. *W. J. Hooker, Eſq.*

---

WE are indebted to Mr. Hooker for the preſent addition to the Britiſh Flora, he having diſcovered it growing plentifully among the moſs on ſhady banks in the neighbourhood of Norwich, and by comparing it with authentic ſpecimens proved that it is the C. Acharii of the above mentioned German authors. It forms ſtrata ſeveral inches in circumference of an olive brown color, by which it may be at once diſtinguiſhed from C. velutina, which it moſt reſembles in its mode and place of growth. The filaments grow nearly erect and matted together; their nature is rather brittle, and each has rarely more than one branch, which is ſhort, patent and very obtuſe. The length of the joints is nearly equal to double the diameter. The fructification has not been diſcovered.

C. Acharii may be diſtinguiſhed from C. othotrichi by its far different color; by its place and mode of growth; by its filaments which are much leſs branched, and by its ſhorter joints.

In drying it adheres to neither Glaſs or Paper.

A. C. Acharii, natural ſize.
B. Ditto, magnified 3.
C. Ditto, ditto 1.

A

# CONFERVA HOOKERI.

C. filamentis primariis inarticulatis, ramulis pinnatis, tenuibus, flexuofis, undique fparfis, pallide rubro-fufcefcentibus; pinnulis alternis articulatis; articulis diametro fefquilongioribus.

On Rocks in the Sea at Cawfie, Murrayfhire; *Mr. Hooker* and *Mr Borrer* Holyhead; *Rev. Hugh Davies.* Bantry Bay; *Mifs Hutchins.*

I HAVE been favored with a fpecimen of this plant gathered by the Rev. Hugh Davies, and which was marked by Hudfon " C. *albida*." The fpecimen in the Dillenian Herbarium according with Hudfon's reference is however very different, and agrees better with both the name and defcription of that fpecies.

Mr. Hooker favored me with the prefent drawing from a fpecimen which he gathered during his late tour through Scotland, and I have a pleafure in embracing the opportunity it affords me of thus acknowledging the great affiftance which I have received from him, by diftinguifhing it with his name.

*C. Hookeri* grows to the length of two or three inches, and whilft recent has a remarkably gelatinous appearance. The color is a pale reddifh brown. The principal ftems are entirely deftitute of diffepiments, and are of an unequal thicknefs, fo that if examined feparately they might be miftaken for an Ulva: they are however befet with pinnated, flexuofe, jointed ramuli, and which are remarkably flender in proportion to the thicknefs of the ftem: the pinnulæ are alternate. The length of the joints is about half greater than the diameter. The capfules are nearly globular and of the fame nature with those of C. *rofea.* It is in fize and mode of growth nearly allied to C. *arbufcula,* but differs in its color, in being branched throughout its whole length, in having pinnated inftead of multifid ramuli, and in the fhape and difpofition of its capfules.

In drying it adheres ſirmly to both Glaſs and Paper.

| | | | |
|---|---|---|---|
| A. | C. Hookeri, | natural ſize. | |
| B. | Ditto, | magnified | 3. |
| C. | Ditto, | ditto | 2. |
| D. | Ditto, | ditto | 1. |

*Conferva Brodiæi.*

# CONFERVA BRODIÆI.

C. filamentis ramoſiſſimis venoſis purpureo-nigreſcentibus; ramis elongatis, ramulis ſparſis, patentibus, faſciculatis, multifidis; articulis ramorum obſoletis, ramulorum diametro ſub-longioribus.

Rocks in the Sea. Near Forres; *James Brodie, Eſq.* Bantry Bay; *Miſs Hutchins.* At Falmouth; *Mr. Turner.* At Seaton, Devon; *Mr. Griffiths.* Sometimes thrown on the ſhore at Dover.

---

OF this ſpecies I firſt received ſpecimens from Mr. Brodie, and have named it after him as an acknowledgement of the kind attention with which he has honored me, and of the aſſiſtance which he has given to this work.

C. Brodiæi is among the moſt magnificent of the genus, often extending to a foot and a half or two feet in length, and puſhing forth from a diſcoid baſe ſeveral main filaments as thick as ſmall twine and of a blackiſh purple color. Theſe are beſet with ſcattered branches of uncertain length, which ariſe in a direction between horizontal and patent: along the branches at irregular intervals cluſters of ſlender ramuli are diſpoſed, from a quarter to half an inch long, multifid in a ſub dichotomous manner, and acuminated at their apices. The whole of the branches and ramuli are of rich deep red-brown color when freſh, but turn black on drying, and are always ſtrongly marked with dark longitudinal veins. The capſules are ovate, ſeſſile, and plentifully ſcattered over the ultimate ramuli, ſometimes on their ſides, and ſometimes at the axillæ of the diviſions. Beſides theſe C. Brodiæi, in common with moſt other of the marine ſpecies, preſents what is uſually conſidered as another kind of fructification, conſiſting of ſphærical globules imbedded in the ultimate ramuli, but of their real nature I confeſs that I am unable to ſatisfy myſelf.

16

The whole of the plant is remarkably thick and bufhy, and its mode of growth flexuofe, by which, with its peculiar color, it may be readily known from its congeners. Neither the main ftem or principal branches fhew any appearance of diffepiments, but in the ramuli they are very ftriking, and divide them into joints whofe length and diameter are nearly equal.

The drawing here reprefented was made by Mifs Hutchins from a frefh plant, and by her communicated to my friend Mr. Turner.

In drying it adheres but very flightly to Paper and not at all to Glafs.

A. C. Brodiæi, natural fize.
B. Ditto magnified 4.
C. Ditto ditto 3.

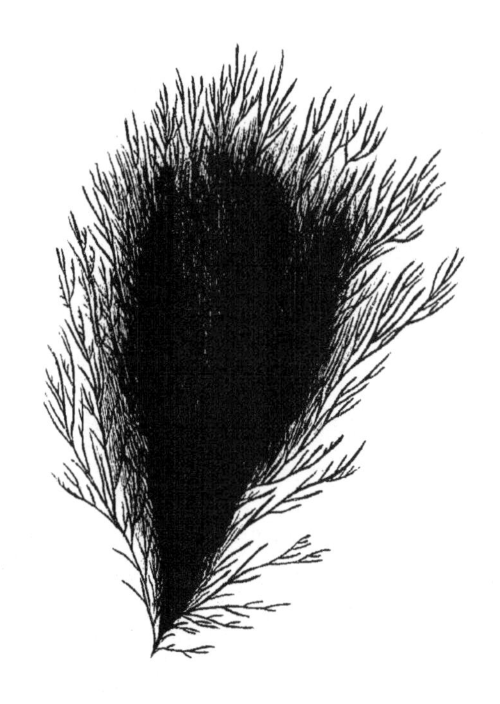

# CONFERVA HUTCHINSIÆ.

C. filamentis ramofiffimis, flexuofis, fub cartilagineis, fragilibus, glauco-viridibus; ramis fparfis; ramulis fub fecundis erectis, articulis torulofis, diametro duplo logioribus.

In Bantry Bay, not rare. *Mifs Hutchins.*

---

I HAVE feen no fpecimen of this beautiful and ftriking fpecies befides what I have received through the favor of my friend Mr. Turner from Mifs Hutchins, by whofe name I have had a peculiar pleafure in calling it, as I know few, if any Botanifts, whofe zeal and fuccefs in the purfuit of Natural Hiftory better deferve fuch a compliment. I am alfo indebted to her for the drawing here reprefented.

The color, according to Mifs Hutchins, is a beautiful glaucous green, with changeable tints when frefh, and under the water appears almoft white. The fubftance is rather ftiff and approaches to cartilaginous. The root is a largifh difk, giving rife to numerous cluftered filaments from three to eight inches long, fomewhat thicker than horfe hair, of equal fize from bafe to fummit, flexuofe, very much and irregularly branched; branches between erect and patent, loofely befet with others difpofed in the fame irregular manner, and thefe again with others; the ultimate ones are fhort, moftly fimple, generally placed more on one fide of the branch than on the other, and very flightly attenuated towards the apices. The length of the joints is uncertain even upon the fame filament, but is about twice greater than the width; in the middle they are flightly torulofe. No fruit has yet been difcovered. It is however proper to obferve that this defcription has been made with Mr. Turner's affiftance from dried fpecimens, as I have not been able to obtain the plant in any other ftate.

C. Hutchinſiæ approaches moſt nearly to C. diffuſa, from which it differs in the greater ſize of its filaments and in the much ſhorter joints, which are not as in that ſpecies regularly cylindrical but conſtantly ſwollen in the center. The ſame characters, and ſtill more its flexuoſe mode of growth, diſtinguiſh it at firſt ſight from C. rupeſtris.

In drying it adheres ſlightly to either Glaſs or Paper.

A. C. Hutchinſiæ, natural ſize.

B. Ditto, magnified 3.

# CONFERVA PEDICELLATA.

C. filamentis dichotomo-ramosis, diffusis, rubris; ramulis alternis, multifidis, apicibus furcatis; articulis sursum incrassatis, diametro sub-quintuplo longioribus.

C. pedicellata. Eng. Bot. t. 1817. (malè).

On the Beach at Selsey and Brighton; *Mr. Borrer*. Bantry Bay; *Miss Hutchins*.

---

WE are indebted to Mr. Botrer for this elegant addition to the British Flora, he having first discovered it in Sussex, and I am not aware of its having been found by any other botanist except Miss Hutchins, to whom I am obliged, through Mr. Turner, for the present drawing, made by herself from recent specimens which she gathered on the shore at Bantry.

C. pedicellata grows about four inches in length, and is of a deep red inclining to rose color. The filaments are repeatedly divided with rather diffuse and dichotomous branches: the ramuli are alternate, and somewhat fasciculated with forked apices; in the specimens from Miss Hutchins they are obtuse in every part of the plant, and so are the lower ones of those from Mr. Borrer, but in these latter the uppermost are elongated and gradually attenuated towards their summits. The length of the joints is rather variable, but mostly about five times greater than the diameter, and excepting those which constitute the terminations of the ramuli, they are always thickest at the apices. The capsules are on short fruit stalks, joined at the base, ovate, solitary, and most frequently placed in the upper forks of the ramuli.

In drying it adheres to both Olass and Paper.

| | | |
|---|---|---|
| A. | C. pedicellata, | natural fize. |
| B. | Ditto, | magnified 3. |
| C. | Ditto, | ditto |

FINIS.

W. Phillips, Printer,
George Yard, Lombard Street, London.

decorticans

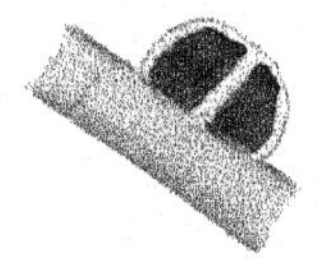

confervicola

C scopul

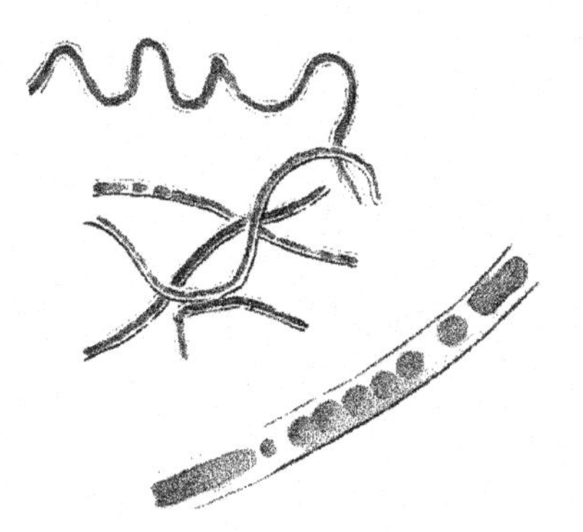

C majuscula

C distorta

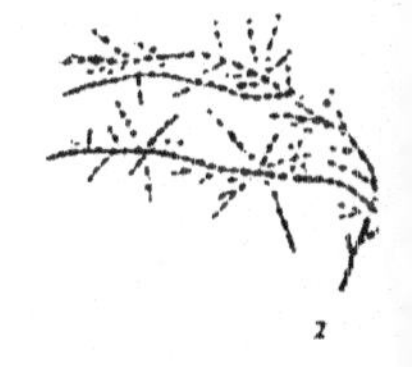

2

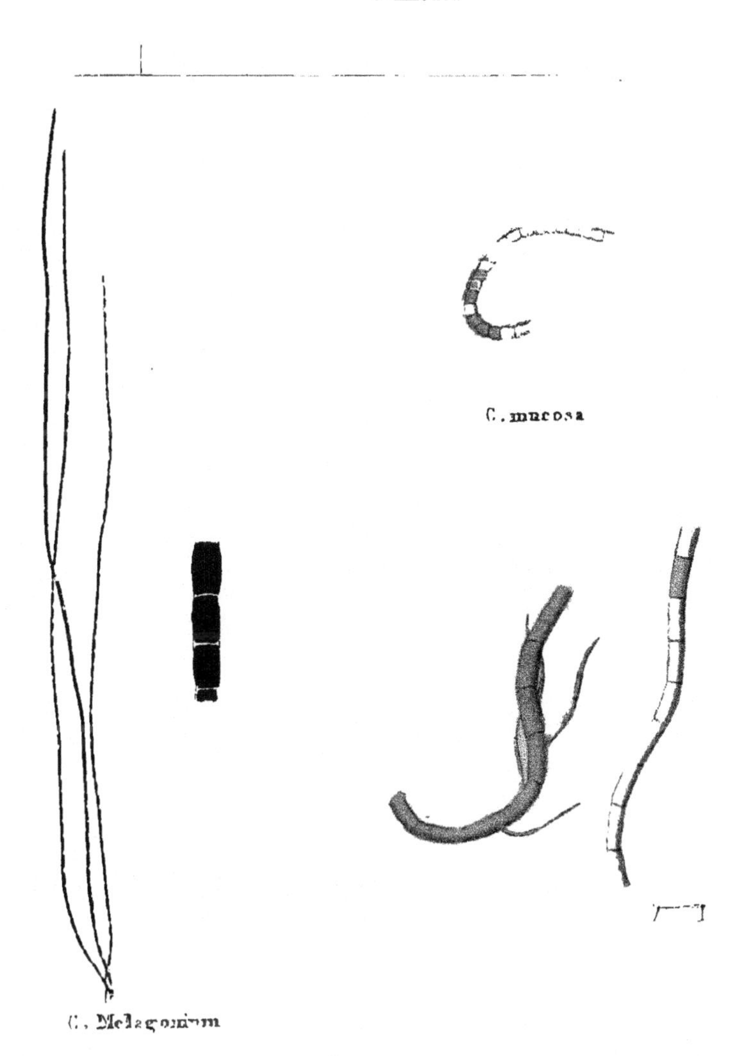
C. lineata
C. mucosa
C. Melagonium

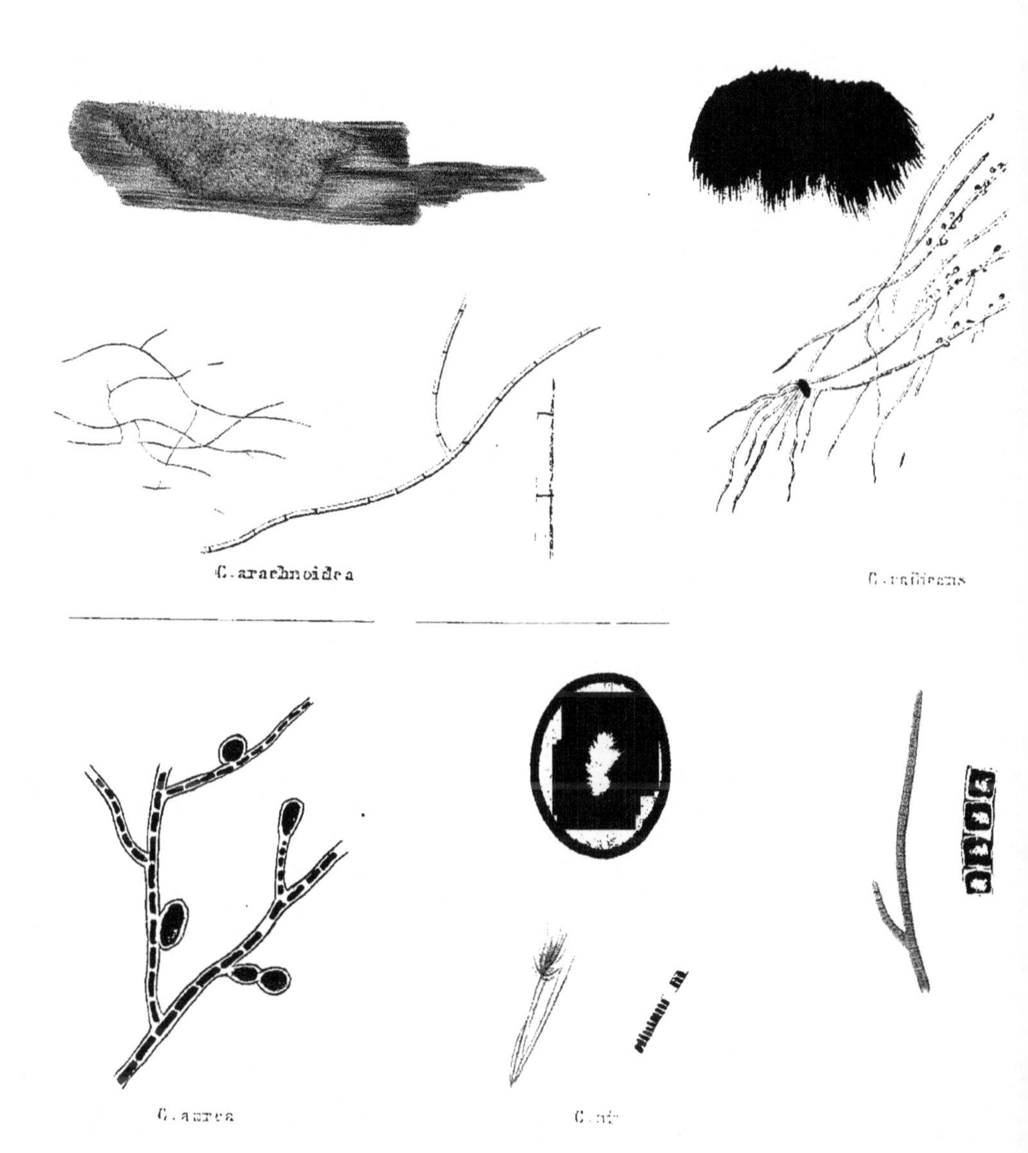
C. arachnoidea
C. aurea

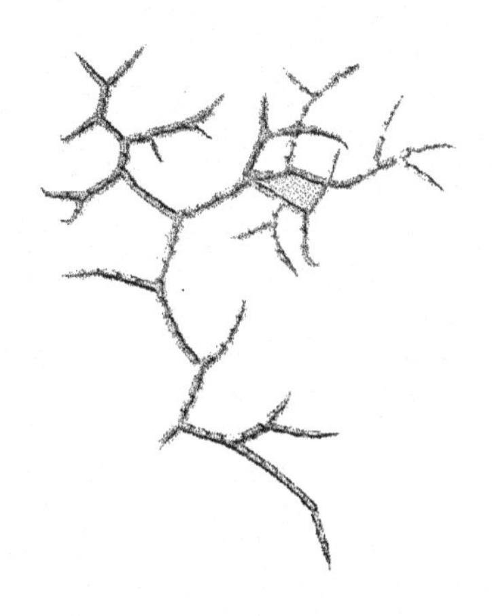

C. cryptarum

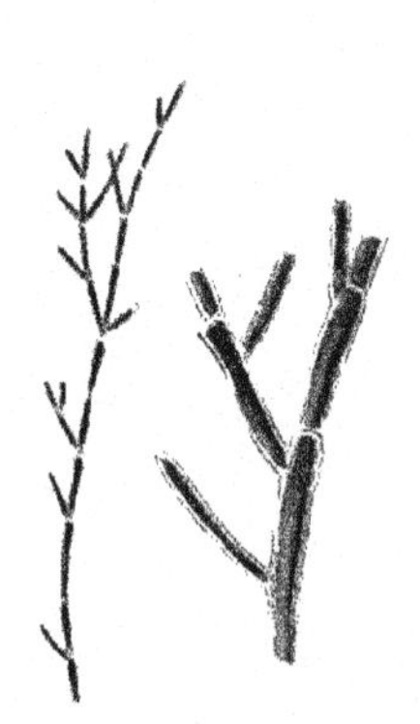

C. Brownii

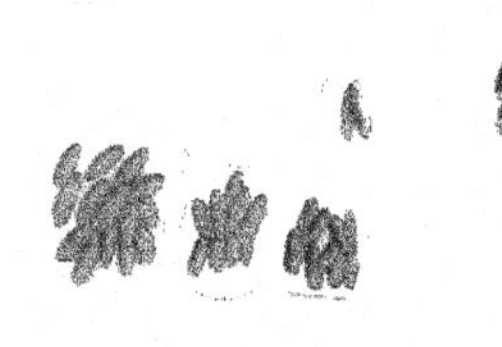

C. multicapsularis

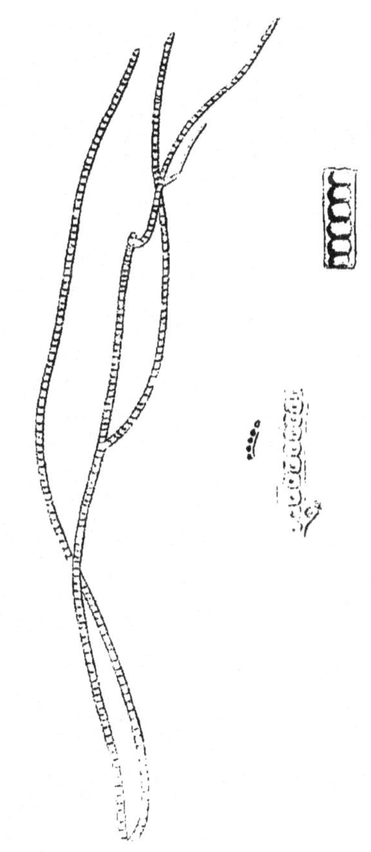

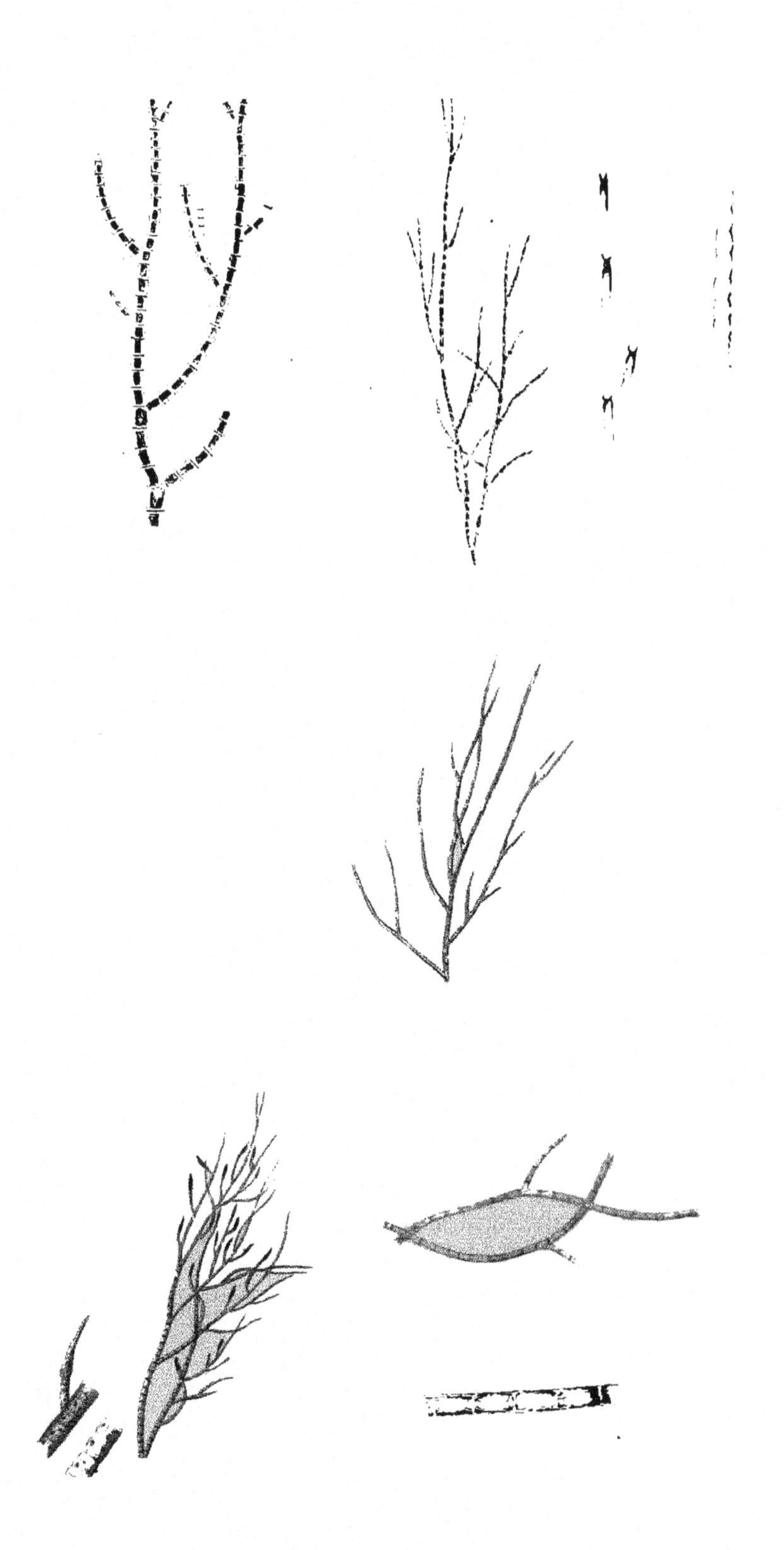

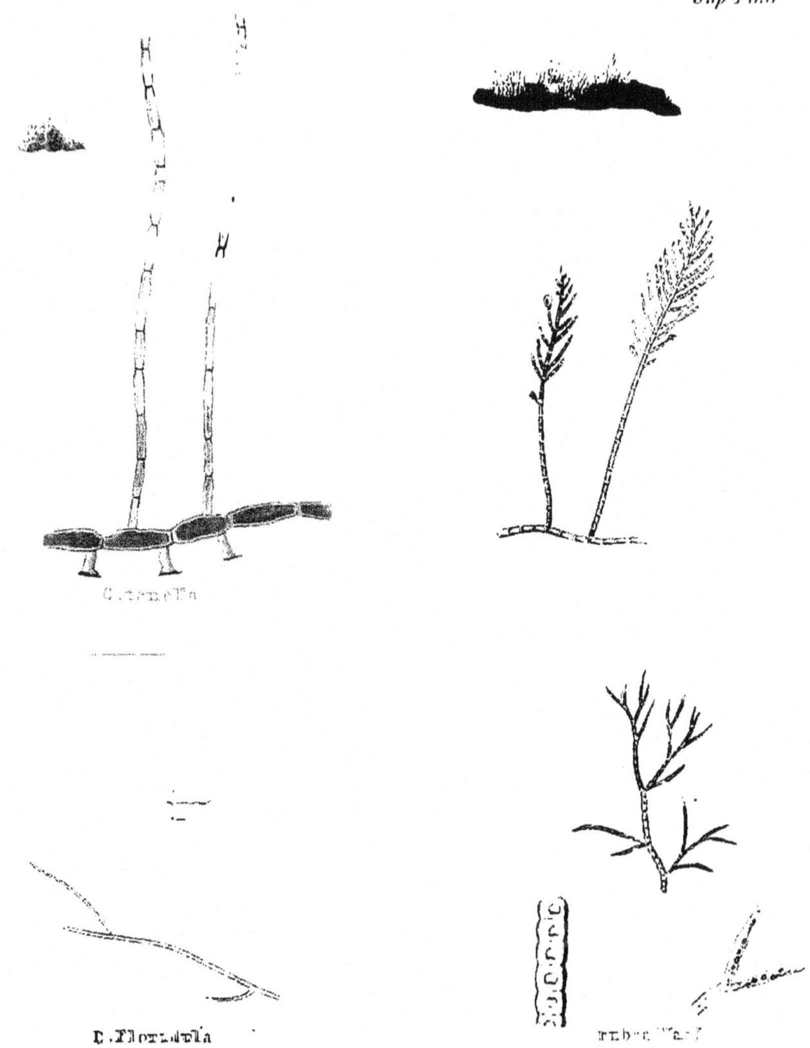
Sup Plate
D. Floridula

[illegible]. patens

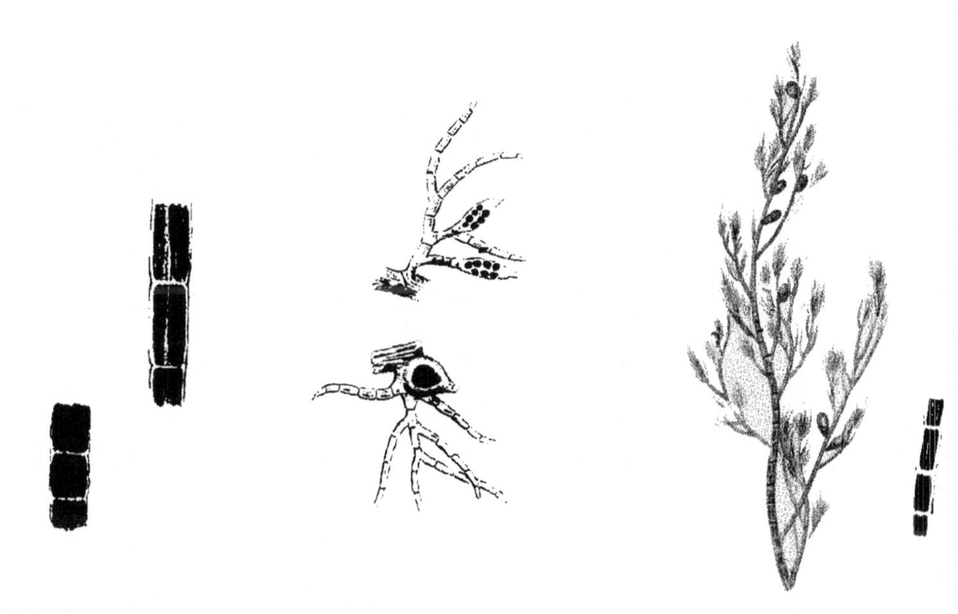

plata

# INDEX.

The numbers in the firſt column refer to the pages of the Introduction and Synopſis; in the ſecond to the Plate and correſponding Deſcription.

The names printed in italics are ſynonyms.

# INDEX.

# ERRATA.

Page 32, line 18, for T. 32 read T. 56.
—— 37, —— 2, for T. 43 read T. 41.
—— 41, ——25, for *lubrica* read *lucens*.
—— 50, ——27, for VAUCHER read *Conj. stellina*. VAUCHER. *Hist. des Conferves*, p. 75. t. 7. f. 1.
—— 61, ——23, for T. 89 read T. 69.
—— 79, ——20, after T. add 100.
—— 81, ——13, after T. add 107.

The numbers to five Plates in the fifth Fasciculus, were omitted by the Engraver; they ſhould ſtand as follows.

| | |
|---|---|
| C. ſtricta. | T. 40. |
| C. amphibia. | T. 41. |
| C. ſpongioſa. | T. 42. |
| C. purpurea. | T. 43. |
| C. polymorpha. | T. 44. |